SOME ASPECTS OF MOLECULAR INTERACTIONS IN DENSE MEDIA

BY

Dr L. JANSEN

ASSISTANT PROFESSOR, UNIVERSITY OF MARYLAND

SPRINGER-SCIENCE+BUSINESS MEDIA, B.V.

1955

SOME ASPECTS OF MOLECULAR
INTERACTIONS IN DENSE MEDIA

© *Springer Science+Business Media Dordrecht 1955*

Originally published by Martinus Nijhoff, The Hague, Netherlands in 1955

ISBN 978-94-017-5684-6   ISBN 978-94-017-5992-2 (eBook)
DOI 10.1007/978-94-017-5992-2

CONTENTS

Summary   1

Chapter I. On the Theory of Molecular Polarization in Gases (I)
Effects of molecular interactions on the polarizability of spherical nonpolar molecules   4
   § 1. Introduction   4
   § 2. Formalism and notation   6
   § 3. First- and second orders in the dipole moment   8
   § 4. Third order approximation of the dipole moment 10
   § 5. Coupled harmonic oscillators 11
   § 6. Atomic hydrogen 12
   § 7. Helium atoms 13
   § 8. Summary of results and discussion 14
Appendices 15

Chapter II. On the Theory of Molecular Polarization in Gases (II)
Effects of molecular interactions on the Clausius-Mosotti function for systems of spherical nonpolar molecules 19
   § 1. Introduction 19
   § 2. The Clausius-Mosotti formula for dense gases 20
   § 3. The average polarizability as a function of density 22
   § 4. Evaluation of the correction term $R(n,T)$ 24
   § 5. The Clausius-Mosotti function for argon 26
   § 6. Summary and conclusions 27

Chapter III. Deviations from Additivity of the Intermolecular Field at High Densities
   § 1. Introduction 30
   § 2. Formalism 33
   § 3. Second-order forces 34
   § 4. Thermodynamic functions 35
   § 5. Relative magnitude of effects 36
   § 6. Discussion 37

CHAPTER IV. ON INTERMOLECULAR FORCES AND CRYSTAL STRUCTURES OF THE RARE GASES
§ 1. Introduction 40
§ 2. Determination of zero-point energy. 42
§ 3. The crystal energies at absolute zero 43
§ 4. Results for the Lennard Jones (s,6) potential 45
§ 5. Results for the modified Buckingham potential (a,6) . . . 46
§ 6. Discussion of the results . . . . 4.

CHAPTER V. CONTRIBUTION TO THE ANALYSIS OF MOLECULAR IN-TERACTIONS IN COMPRESSED NITROGEN AND CARBON MONOXIDE
Part I. *The molecular field in the high density gas state* 49
§ 1. Introduction 49
§ 2. Thermodynamic properties and intermolecular parameters of nitrogen and carbon monoxide. 50
§ 3. *a.* Cell theory and the theorem of corresponding states . . 52
§ 3. *b.* Application to nitrogen and carbon monoxide 56
§ 4. *a.* The orientation effect of the dipoles of carbon monoxide at high densities 57
§ 4. *b.* The dipole induction effect at high densities for carbon mo-noxide 59
§ 5. Anisotropy of the dispersion forces 60
§ 6. Influence of the change in polarizability with pressure on the thermodynamic properties of compressed gases 62
§ 7. Summary of results 67
Part II. *Quadrupole moments and properties of the solid states* 68
§ 1. Introduction 68
§ 2. The crystal structures of nitrogen and carbon monoxide . . 69
§ 3. The crystal at absolute zero. 70
§ 4. Orientation and induction effects at 0°K. 72
§ 5. Anisotropy effect in London dispersion forces 74
§ 6. Evaluation of the quadrupole moments 75
Part III. *Quadrupole orientation and induction effects at high tempera-tures* 77
§ 1. Introduction 77
§ 2. Quadrupole orientation interaction 78
§ 3. Quadrupole induction effect 79
§ 4. Summary of results. 81
APPENDICES 81

CHAPTER VI. SOME CONSIDERATIONS ON THE THEORY OF TRANSITIONS IN MOLECULAR CRYSTALS 83
CURRICULUM VITAE 86

SUMMARY

This thesis deals with a few special problems involving interactions between simple, nonpolar atoms or molecules.

In Chapter I the theory of molecular polarization in gases is developed on a quantum mechanical basis for spherical, nonpolar atoms. This permits the calculation of the effect of molecular interactions on the polarizability of the atoms, which is usually considered constant in the statistical theory of the dielectric constant. The interaction Hamiltonian is restricted to the induced dipole term and the polarizability is obtained from perturbation theory as a series in powers of the dimensionless quantity $\alpha_0 \, \mathsf{T}$, where α_0 is the polarizability of a free atom and T stands for the tensor characteristic for dipole interactions (i.e. $\mathsf{T}_{ik} = \boldsymbol{\nabla}_i \boldsymbol{\nabla}_k \, (1/r_{ik})$; r_{ik} is the distance between the centers of atoms i and k). As a special case it is shown that in a system of isotropic harmonic oscillators, interacting through induced dipole forces, the polarizability remains constant in any order of approximation. For hydrogen and helium atoms, however, the polarizability increases in the order T^2 due to interactions between pairs of atoms. This effect gives rise to a deviation from the Clausius-Mosotti expression.

In Chapter II the results obtained in Chapter I are applied to the evaluation of the Clausius-Mosotti function for compressed gases in the density region up to 200 Amagat units. The Yvon-Kirkwood statistical theory is modified to include the effect of interactions on the polarizability. The result is a series expansion of the Clausius-Mosotti function in powers of the density, the coefficients of which can be expressed in terms of the intermolecular forces. It is shown that for helium atoms the effect of interactions on the polarizability gives a contribution to the Clausius-Mosotti function which is of the same order of magnitude and of the same sign as the effect due to translational fluctuations in the induced dipole moments. The results may also be expected to give good agreement with experimental data for argon.

The intermolecular field of an assembly of neutral atoms or molecules is usually assumed to be additive with respect to isolated pairs. In Chapter III the validity of this assumption is analyzed. First a brief review is given of calculations by P. R o s e n (first-order interactions) and by A x i l r o d and T e l l e r (third-order interactions) referring to nonadditive contributions to the intermolecular field. For the evaluation of second-order interactions between molecules at high densities the zero-order wave function must be made antisymmetric with respect to nearest neighbors. It is shown that this correction results in a decrease of the second-order interactions between the molecules, compared with isolated pairs, and that this correction may be interpreted as a "screening effect". The method is based on the model of the "caged" atom or molecule.

In Chapter IV the relation between the intermolecular potential field and the stability of crystal structures of rare gases is investigated for a face centered cubic lattice and a crystal of hexagonal closest packing. Two forms of the potential field are used: a Lennard Jones $(s, 6)$ potential (s between 7 and 16) and a "modified" Buckingham potential $(a, 6)$. The effect of zero-point energy is taken into account on the basis of Corner's method. It is found that for an additive Lennard Jones potential the hexagonal lattice is somewhat more stable than the cubic structure. For an additive Buckingham potential, with a varying between 10 and 16, the hexagonal lattice is again more stable. For both forms of the interaction considered, the differences in energy between the two structures are of the order of one tenthousandth of the cohesive energy. These results agree with the calculations by K i h a r a and K o b a, who neglected the influence of zero-point energy. It is concluded that neither a Lennard Jones nor a "modified" Buckingham potential can account for the observed crystal structures of the heavy rare gases. The possible importance of many-body forces is discussed.

In Chapter V various physical properties of systems consisting of diatomic molecules are analyzed in three parts. In the first part the thermodynamic properties of compressed nitrogen and carbon monoxide are compared up to a density of 600 Amagat and at temperatures between 0°C and 150°C. The differences between the two gases are discussed on the basis of a detailed study of the intermolecular field. In part II an analysis of the interaction field of solid nitrogen and carbon monoxide is based on experimental values for the sublimation energies of the crystals, extrapolated to 0°K, and on other crystal properties. As in part I, the intermolecular field is assumed to consist of a Lennard Jones $(6, 12)$ part, a term due to orientation and induction effects between the permanent dipoles (CO) and permanent quadrupoles (CO and N_2), plus a term for the effect of anisotropy in the London forces. The orientation and induction effects are evaluated for four shells of molecules around a central molecule. The values for the quadrupole moments calculated in this way are found to agree well with those determined recently from microwave spectra. With these values of the quadrupole moments the analysis of molecular interactions in compressed nitrogen and carbon monoxide is concluded in part III. It is found that neither the quadrupole orientation, nor quadrupole induction or dipole-quadrupole orientation energies contribute significantly to the thermodynamic properties of these gases at high

2

temperatures (between 0°C and 150°C). The calculations are followed by a summary of the main results of the analysis.

In Chapter VI the transitions observed in crystals of nitrogen and carbon monoxide are discussed; a theoretical outline is given of an analysis to explain this phenomenon.

The research on which this thesis is based was carried out at the Molecular Physics Laboratory of the University of Maryland, College Park, U.S.A. This laboratory is supported financially by the U.S. Office of Naval Research under Contract Nonr-595(02).

L. JANSEN.

Present address:
    Institute of Molecular Physics
    University of Maryland
    College Park, Md,
    U.S.A.

Chapter I

ON THE THEORY OF MOLECULAR POLARIZATION IN GASES

I. EFFECT OF MOLECULAR INTERACTIONS ON THE POLARIZABILITY OF SPHERICAL NONPOLAR MOLECULES

§ 1. *Introduction*. The validity of the Clausius-Mosotti equation for the relation between the polarizability a of a molecule and the dielectric constant ε, viz.

$$\frac{\varepsilon - 1}{\varepsilon + 2} = \tfrac{4}{3} \pi N a \qquad (1)$$

has been the subject of many experimental and theoretical investigations. The volume per mole is V and N is Avogrado's number. It is well known that polar substances show large deviations from (1). Slight deviations have also been found for nonpolar gases under high pressures and for nonpolar liquids. U h l i g, K i r k w o o d and K e y e s [1]) observed an increase of $(\varepsilon-1)V/(\varepsilon + 2)$ for compressed CO_2 up to 200 atm. The experiments for the same gas by M i c h e l s and M i c h e l s [2]) showed that at higher densities (corresponding to pressures of about 1000 atm.) the Clausius-Mosotti function decreases with increasing pressure. These experiments were later repeated with increased accuracy by M i c h e l s and K l e e r e k o p e r [3]); the result is that $(\varepsilon - 1) V/(\varepsilon + 2)$ first increases with density, goes through a maximum at about 200 atm. and then decreases. For monoatomic molecules only a few data are available. M i c h e l s, t e n S e l d a m and O v e r d ij k [4]) measured the change in the Clausius-Mosotti function with density for argon up to a density of about 500 Amagat, and M i c h e l s and B o t z e n [5]) performed analogous experiments for the Lorentz-Lorenz expression $(n^2 - 1) V/(n^2 + 2)$. The same general results are obtained: the Clausius-Mosotti and Lorentz-Lorenz functions first increase slightly with increasing pressure, and then decrease.

Several theoretical explanations have been offered for these deviations. The usual derivation of (1) is based on Lorentz' formula for the mean electric field intensity acting on a molecule [6]). It is well known that Lorentz' method

contains approximations which fail in the case of polar liquids (see e.g. v a n
V l e c k [7])). A number of attempts has been made to improve the Clausius-
Mosotti equation. Y v o n [8]), v a n V l e c k [7]), K i r k w o o d [9]) and later B o e t t-
c h e r [10]) derived formulae which give better agreement with experiments.
Whereas the calculations by the first three authors are based on molecular
theory and lead to a series expression for $(\varepsilon-1)\ V/(\varepsilon+2)$, Boettcher obtains
a closed expression based on the Onsager model, in which one molecule is
treated explicity and the other molecules are replaced by a continuum. A
critical analysis of the different theories and a reformulation of the Yvon-
Kirkwood statistical method was presented by F u l l e r B r o w n [11]).

The statistical approximation inherent in the Lorentz derivation of (1) can
best be understood on the basis of the Yvon-Kirkwood theory. The Lorentz
formula assumes that the mean moment of a molecule known to be at a
specified position with respect to another molecule is the same as the mean
moment of a molecule without such specification. While this is certainly an
excellent approximation in crystals, it cannot be accepted for fluids and
compressed gases because of appreciable fluctuations in the local densities
of the system. For an extensive discussion of this point we must refer to the
papers by K i r k w o o d, Y v o n, and to the analysis by F u l l e r
B r o w n. The effect of statistical fluctuations is that the Clausius-Mosotti
function first increases with increasing density, goes through a maximum
and then decreases, in qualitative agreement with the experiments. The
series expansion can be formally written as

$$\frac{\varepsilon - 1}{\varepsilon + 2}\ V = \tfrac{4}{3}\pi\ N a(1 + S) \tag{2}$$

following a slight modification of the Yvon-Kirkwood theory by d e B o e r,
v a n d e r M a e s e n and t e n S e l d a m [12]). The quantity S is a series
in powers of the density with coefficients which can be expressed in terms of
the polarizability of the molecules and of the intermolecular forces. If there
were no density fluctuations, then S would be zero and (2) becomes identical
with the Clausius-Mosotti expression (1). A quantitative comparison with
experiments for monoatomic gases, in this case for argon, was made by d e
B o e r, v a n d e r M a e s e n and t e n S e l d a m [12]). The intermolecular
field (which figures in the statistical theory in the molecular distribution
functions) was taken to be of the Lennard Jones form; calculations were also
carried out for the Herzfeld "square well" potential function. Qualitatively
there is good agreement but at low densities the experimental curve seems to
increase more rapidly than the fluctuation theory predicts, whereas at high
densities the experimental decrease in $(\varepsilon - 1)\ V/(\varepsilon + 2)$ is somewhat more
rapid with increasing density than the theoretical curve indicates.

In all the theories mentioned above, the polarizability was considered
independent of the density. A more complete theory should therefore include

the effect of molecular interactions on the polarizability of the molecules. It can be shown [13] [14] that first-order interactions (i.e. repulsive forces) between neutral molecules result in a decrease of the polarizability, but this effect fails to explain the deviations at low densities.

It is the purpose of this paper to present a low-density calculation of the change in the polarizability of spherical, nonpolar molecules with density, on the basis of a perturbation expansion. It will be shown that one thus can account in principle for the initial increase of the polarizability with decreasing intermolecular distances. The calculations will be restricted to intermolecular distances which are so large that only the dipole term in the multipole expansion of the Coulomb interactions need to be considered, and that exchange forces may be neglected.

§ 2. *Formalism and notation.* The polarizability α_i of a molecule i in a system of N interacting identical molecules is given by *) **)

$$\langle \mathbf{p}_i \rangle = \alpha_i \cdot (\mathbf{E}_0 - \Sigma_{k \neq i} \mathsf{T}_{ik} \cdot \langle \mathbf{p}_k \rangle). \tag{3}$$

In general α_i is a tensor quantity. Further, $\langle \mathbf{p}_i \rangle$ stands for the quantum mechanical average of the dipole moment $\mathbf{p}_i$ of molecule i, induced by the external static electric field $\mathbf{E}_0$ and by the field $- \Sigma_{k \neq i} \mathsf{T}_{ik} \cdot \langle \mathbf{p}_k \rangle$ of the induced dipole moments $\mathbf{p}_k$ of the other molecules k. The tensor T_{ik}, characteristic for dipole-dipole interactions is given in dyadic notation by

$$\mathsf{T}_{ik} = \nabla_i \nabla_k (1/r_{ik}) = r_{ik}^{-3} (\mathsf{U} - (3\mathbf{r}_{ik} \mathbf{r}_{ik}/r_{ik}^2)). \tag{4}$$

U is the unit tensor. For our calculations it is convenient to use tensors β_i; $i = 1, 2, \ldots, N$, which are defined by

$$\langle \mathbf{p}_i \rangle = \beta_i \cdot \mathbf{E}_0 \tag{5}$$

so that for noninteracting molecules we have $\beta_i = a_0 \mathsf{U}$ (where a_0 is the polarizability of a free molecule. The polarizability α_i can then be expressed in terms of the β_i and T_{ik} with the help of (3) and (5)

$$\beta_i = \alpha_i \cdot (\mathsf{U} - \Sigma_{k \neq i} \mathsf{T}_{ik} \cdot \beta_k),$$

where U is again the unit tensor, or:

$$\alpha_i = \beta_i \cdot (\mathsf{U} - \Sigma_{k \neq i} \mathsf{T}_{ik} \cdot \beta_k)^{-1}. \tag{6}$$

This expression can be expanded in powers of $\mathsf{T} . \beta$

$$\alpha_i = \beta_i + \beta_i \cdot \Sigma_{k \neq i} \mathsf{T}_{ik} \cdot \beta_k + \beta_i \cdot \Sigma_{k \neq i} \Sigma_{l \neq i} \mathsf{T}_{ik} \cdot \beta_k \cdot \mathsf{T}_{il} \cdot \beta_l + \ldots \tag{7}$$

---

*) See Appendix I on tensor notation.

**) The quantity α_i has been introduced by analogy with the case of particles with constant polarizability. That such a quantity can be defined according to (3) is justified by the perturbation theory of §§ 3 and 4.

We will calculate $\langle \mathbf{p}_i \rangle$ or $\boldsymbol{\beta}_i$ in successive orders of approximation with perturbation theory; the zeroth order refers to the free molecules in the absence of a field. Eq. (7) then establishes the relation between $\boldsymbol{\beta}_i$, $\boldsymbol{\beta}_k$, ..., T_{ik} and $\boldsymbol{\alpha}_i$. The evaluation of $\langle \mathbf{p}_i \rangle$ is restricted to neutral, nonpolar molecules which are optically isotropic, and have nondegenerate ground states. The system of N such molecules is placed in a static external electric field $\mathbf{E}_0$ and the molecules interact through induced dipole forces (i.e. the leading term in the multipole expansion of the Coulomb interactions; at higher densities the interactions cannot be restricted to dipoles only but must include poles of higher order). The total Hamiltonian of the system is

$$H = H_0 + H',\qquad\qquad(8)$$

with

$$H' = -\Sigma_i\, \mathbf{p}_i \cdot \mathbf{E}_0 + \tfrac{1}{2}\Sigma_i \Sigma_{k \neq i}\, \mathbf{p}_i \cdot T_{ik} \cdot \mathbf{p}_k.$$

The Hamiltonian of a system of free molecules in the absence of the external field is H_0; T_{ik} is the tensor defined in (4). The dipole moment operator $\mathbf{p}_i$ of molecule i is given by $\mathbf{p}_i = \Sigma_n\, e_n \mathbf{r}_n$, where n numbers the charges in the molecule and $\mathbf{r}_n$ is the radius vector of the n-th charge with respect to an arbitrary origin.

It is assumed that the system of N interacting molecules in the external field $\mathbf{E}_0$ is in its ground state, specified by Ψ_0, an eigenfunction of the total Hamiltonian H. The energy eigenfunctions of the unperturbed system are $\varphi_\varkappa$; ($\varkappa = 0, 1, \ldots$) where $\varkappa$ labels the various eigenstates. The quantum mechanical average of the dipole moment $\mathbf{p}_i$ is given by

$$\langle \mathbf{p}_i \rangle = \int \Psi_0^* \, \mathbf{p}_i \, \Psi_0 \, d\tau\qquad\qquad(9)$$

as a $3N$-dimensional integral over configuration space. The perturbed wave function Ψ is expanded in successive orders of approximation as

$$\Psi = \Psi^{(0)} + \Psi^{(1)} + \Psi^{(2)} + \ldots\ldots\qquad\qquad(10)$$

The zero-, first-, etc. orders will be written in "bra-ket" notation as $|0\rangle$, $|1\rangle$, etc. The corresponding expansion for $\langle \mathbf{p}_i \rangle$ is then:

$$\langle \mathbf{p}_i \rangle = \langle \mathbf{p}_i \rangle^{(0)} + \langle \mathbf{p}_i \rangle^{(1)} + \langle \mathbf{p}_i \rangle^{(2)} + \langle \mathbf{p}_i \rangle^{(3)},\qquad\qquad(11)$$

with

$$\langle \mathbf{p}_i \rangle^{(0)} = \langle 0|\mathbf{p}_i|0\rangle = 0,$$

$$\langle \mathbf{p}_i \rangle^{(1)} = \langle 0|\mathbf{p}_i| 1\rangle + \langle 1|\mathbf{p}_i| 0\rangle = 2\langle 0|\mathbf{p}_i| 1\rangle,\qquad\qquad(12)$$

$$\langle \mathbf{p}_i \rangle^{(2)} = 2\langle 0|\mathbf{p}_i| 2\rangle + \langle 1|\mathbf{p}_i| 1\rangle,\qquad\qquad(13)$$

$$\langle \mathbf{p}_i \rangle^{(3)} = 2\langle 0|\mathbf{p}_i| 3\rangle + 2\langle 1|\mathbf{p}_i| 2\rangle,\ \text{etc.},\qquad\qquad(14)$$

since $\mathbf{p}_i$ is a Hermitean operator. Using the perturbation expressions for the different order wave functions [15] and noting that $(H')_{00} = 0$, the following expressions are obtained

$$\Psi_0^{(0)} = \varphi_0,$$

$$\Psi_0^{(1)} = \Sigma_{\varkappa \neq 0} \frac{\varphi_\varkappa H'_{\varkappa 0}}{E_0 - E_\varkappa}, \tag{15}$$

$$\Psi_0^{(2)} = \Sigma_{\varkappa \neq 0} \Sigma_{\lambda \neq 0} \frac{\varphi_\varkappa H'_{\varkappa \lambda} H'_{\lambda 0}}{(E_0 - E_\varkappa)(E_0 - E_\lambda)} - \tfrac{1}{2} \Sigma_{\varkappa \neq 0} \frac{\varphi_0 H'_{0\varkappa} H'_{\varkappa 0}}{(E_0 - E_\varkappa)^2}, \tag{16}$$

$$\Psi_0^{(3)} = \Sigma_{\varkappa \neq 0} \Sigma_{\lambda \neq 0} \Sigma_{\mu \neq 0} \frac{\varphi_\varkappa H'_{\varkappa \lambda} H'_{\lambda \mu} H'_{\mu 0}}{(E_0 - E_\varkappa)(E_0 - E_\lambda)(E_0 - E_\mu)} -$$

$$- \Sigma_{\varkappa \neq 0} \Sigma_{\lambda \neq 0} \frac{\varphi_\varkappa H'_{\varkappa 0}}{(E_0 - E_\varkappa)^2} \frac{H'_{0\lambda} H'_{\lambda 0}}{(E_0 - E_\lambda)} -$$

$$- \Sigma_{\varkappa \neq 0} \Sigma_{\lambda \neq 0} \frac{\varphi_\varkappa H'_{\varkappa 0}}{(E_0 - E_\varkappa)} \frac{H'_{0\lambda} H'_{\lambda 0}}{(E_0 - E_\lambda)^2} - \Sigma_{\varkappa \neq 0} \Sigma_{\lambda \neq 0} \frac{\varphi_0 H'_{0\varkappa} H'_{\varkappa \lambda} H'_{\lambda 0}}{(E_0 - E_\varkappa)^2 (E_0 - E_\lambda)}, \tag{17}$$

where H' is given by (8). The difference between the energy eigenvalues of the ground state and the $\varkappa$-th excited state of the unperturbed system is $E_0 - E_\varkappa$. All matrix elements are calculated in the system of eigen-functions of H_0.

From (12–14) one obtains explicit expressions for the different orders of $\langle \mathbf{p}_i \rangle$. Since H' contains a contribution linear in $\mathbf{E}_0$ and a part linear in T, $\langle \mathbf{p}_i \rangle$ can be expressed as a power series in $\mathbf{E}_0$ and T. Only the terms linear in $\mathbf{E}_0$ are of interest (i.e., we calculate the polarizability for vanishing fieldstrength $\mathbf{E}_0$). The successive approximations in α_i are *)

$$\alpha_i^{(0)} = \beta_i^{(0)}, \tag{18}$$

$$\alpha_i^{(1)} = \beta_i^{(1)} + \Sigma_{k \neq i} \beta_i^{(0)} \cdot \mathsf{T}_{ik} \cdot \beta_k^{(0)}, \tag{19}$$

$$\alpha_i^{(2)} = \beta_i^{(2)} + \Sigma_{k \neq i} \beta_i^{(1)} \cdot \mathsf{T}_{ik} \cdot \beta_k^{(0)} + \Sigma_{k \neq i} \beta_i^{(0)} \cdot \mathsf{T}_{ik} \cdot \beta_k^{(1)} +$$

$$+ \beta_i^{(0)} \cdot \Sigma_{k \neq i} \Sigma_{l \neq i} \mathsf{T}_{ik} \cdot \beta_k^{(0)} \cdot \mathsf{T}_{il} \cdot \beta_l^{(0)}, \tag{20}$$

etc.

§3. *First- and second orders in the dipole moment.* If the molecules are sufficiently far apart the zero-order wave function of the perturbed system may be written as:

$$\varphi_0 = \Pi_i u_0^i, \tag{21}$$

where u_0^i is the wave function for the ground state of a free molecule i in the absence of interactions and without the external field. For the first-order correction to $\langle \mathbf{p}_i \rangle$ we have, with (12) and (15)

$$\langle \mathbf{p}_i \rangle^{(1)} = 2 \Sigma_{\varkappa \neq 0} \frac{(\mathbf{p}_i)_{0\varkappa} H'_{\varkappa 0}}{(E_0 - E_\varkappa)}.$$

---

*) The orders of approximation in β_i and α_i are lowered by one, compared with the orders in $\langle \mathbf{p}_i \rangle$, since otherwise $\alpha_i^{(0)} = 0$, which is meaningless.

8

Only the term $-\mathbf{p}_i \cdot \mathbf{E}_0$ of H' contributes to the expression, since $\langle \mathbf{p}_k \rangle^{(0)} = 0$. Therefore, with (21)

$$\langle \mathbf{p}_i \rangle^{(1)} = -2 \, \Sigma_{\varkappa_i \neq 0} \, \frac{(\mathbf{p}_i)_{0\varkappa_i} \, (\mathbf{p}_i)_{\varkappa_i 0} \cdot \mathbf{E}_0}{(E_{0_i} - E_{\varkappa_i})}, \tag{22}$$

where $\varkappa_i$ labels the excited states of molecule i. The perturbation expression for the polarizability of a free molecule, a_0, is

$$a_0 \, \mathsf{U} = -2 \, \Sigma_{\varkappa_i \neq 0} \, \frac{(\mathbf{p}_i)_{0\varkappa_i} \, (\mathbf{p}_i)_{\varkappa_i 0}}{E_{0_i} - E_{\varkappa_i}} \tag{23}$$

in dyadic notation; U is the unit tensor. From (22) and (23) we have

$$\langle \mathbf{p}_i \rangle^{(1)} = a_0 \mathbf{E}_0, \tag{24}$$

i.e., no effect on the polarizability by induced forces in first order. The second-order approximation $\langle \mathbf{p}_i \rangle^{(2)}$ is obtained from (13), (15), (16).

$$\langle \mathbf{p}_i \rangle^{(2)} = \Sigma_{\varkappa \neq 0} \, \Sigma_{\lambda \neq 0} \, \frac{(\mathbf{p}_i)_{0\varkappa} \, H'_{\varkappa\lambda} \, H'_{\lambda 0}}{(E_0 - E_\varkappa)(E_0 - E_\lambda)} + \Sigma_{\varkappa \neq 0} \, \Sigma_{\lambda \neq 0} \, \frac{H'_{0\varkappa} \, (\mathbf{p}_i)_{\varkappa\lambda} \, H'_{\lambda 0}}{(E_0 - E_\varkappa)(E_0 - E_\lambda)}.$$

For reasons of parity terms quadratic in $\mathbf{E}_0$ vanish, as well as contributions in T^2. The remaining terms are

$$\langle \mathbf{p}_i \rangle^{(2)} = \Sigma_{k \neq i} \left[-2 \, \Sigma_{\varkappa_i \neq 0} \, \Sigma_{\lambda_k \neq 0} \, \frac{(\mathbf{p}_i)_{0\varkappa_i} \, (\mathbf{p}_k \cdot \mathbf{E}_0)_{0\lambda_k} \, (\mathbf{p}_i)_{\varkappa_i 0} \cdot \mathsf{T}_{ik} \cdot (\mathbf{p}_k)_{\lambda_k 0}}{(E_{0_i} - E_{\varkappa_i})(E_{0_i} + E_{0_k} - E_{\varkappa_i} - E_{\lambda_k})} \right.$$

$$-2 \, \Sigma_{\varkappa_i \neq 0} \, \Sigma_{\lambda_k \neq 0} \, \frac{(\mathbf{p}_i)_{0\varkappa_i} \, (\mathbf{p}_i)_{\varkappa_i 0} \cdot \mathsf{T}_{ik} \cdot (\mathbf{p}_k)_{0\lambda_k} \, (\mathbf{p}_k \cdot \mathbf{E}_0)_{\lambda_k 0}}{(E_{0_i} - E_{\varkappa_i})(E_{0_k} - E_{\lambda_k})} -$$

$$- \Sigma_{\varkappa_k \neq 0} \, \Sigma_{\lambda_i \neq 0} \, \frac{(\mathbf{p}_k \cdot \mathbf{E}_0)_{0\varkappa_k} \, (\mathbf{p}_i)_{0\lambda_i} \, (\mathbf{p}_i)_{\lambda_i 0} \cdot \mathsf{T}_{ik} \cdot (\mathbf{p}_k)_{\varkappa_k 0}}{(E_{0_k} - E_{\varkappa_k})(E_{0i} + E_{0_k} - E_{\lambda_i} - E_{\varkappa_k})}$$

$$- \Sigma_{\varkappa_i \neq 0} \, \Sigma_{\varkappa_k \neq 0} \, \frac{(\mathbf{p}_i)_{0\varkappa_i} \cdot \mathsf{T}_{ik} \cdot (\mathbf{p}_k)_{0\varkappa_k} \, (\mathbf{p}_i)_{\varkappa_i 0} \, (\mathbf{p}_k \cdot \mathbf{E}_0)_{\varkappa_k 0}}{(E_{0_i} + E_{0_k} - E_{\varkappa_i} - E_{\varkappa_k})(E_{0_k} - E_{\varkappa_k})}$$

The different contributions can be rearranged and give in total, with eq. (23),

$$\langle \mathbf{p}_i \rangle^{(2)} = -4 \, \Sigma_{k \neq i} \, \Sigma_{\varkappa_i \neq 0} \, \Sigma_{\lambda_k \neq 0} \, \frac{(\mathbf{p}_i)_{0\varkappa_i} \, (\mathbf{p}_i)_{\varkappa_i 0} \cdot \mathsf{T}_{ik} \cdot (\mathbf{p}_k)_{0\lambda_k} \, (\mathbf{p}_k \cdot \mathbf{E}_0)_{\lambda_k 0}}{(E_{0_i} - E_{\varkappa_i})(E_{0_k} - E_{\lambda_k})} = \tag{25}$$

$$= -a_0^2 \, \Sigma_{k \neq i} \, \mathsf{T}_{ik} \cdot \mathbf{E}_0$$

From (24) and (25) we obtain for $\boldsymbol{\beta}_i^{(0)}$ and $\boldsymbol{\beta}_i^{(1)}$

$$\boldsymbol{\beta}_i^{(0)} = a_0 \, \mathsf{U}; \quad \boldsymbol{\beta}_i^{(1)} = -a_0^2 \, \Sigma_{k \neq i} \, \mathsf{T}_{ik}, \tag{26}$$

and, after using (18) and (19)

$$\boldsymbol{\alpha}_i^{(0)} = a_0 \, \mathsf{U}; \quad \boldsymbol{\alpha}_i^{(1)} = -a_0^2 \, \Sigma_{k \neq i} \, \mathsf{T}_{ik} + a_0^2 \, \Sigma_{k \neq i} \, \mathsf{T}_{ik} = 0. \tag{27}$$

Thus far there is no change in the polarizability of spherial, nonpolar molecules by induced dipole interactions. As will be shown below the next higher order change in $\boldsymbol{\alpha}_i$ is in general not zero.

§ 4. *Third order approximation of the dipole moment.* The matrix elements which occur in $\langle \mathbf{p}_i \rangle^{(3)}$ are of the form $(\mathbf{p}_i)_{0\varkappa} (H')_{\varkappa\lambda} (H')_{\lambda\mu} (H')_{\mu 0}$; they reflect interactions between triplets of molecules and contain also pair interactions. Contributions in T^3 are zero, and terms in E_0^3 are not considered. With the help of the perturbed wave functions (15), (16) and (17) we obtain

$$\langle \mathbf{p}_i \rangle^{(3)} = 2 \Sigma_{\varkappa \neq 0} \Sigma_{\lambda \neq 0} \Sigma_{\mu \neq 0} \frac{(\mathbf{p}_i)_{0\varkappa} (H')_{\varkappa\lambda} (H')_{\lambda\mu} (H')_{\mu 0}}{(E_0 - E_\varkappa)(E_0 - E_\lambda)(E_0 - E_\mu)}$$

$$+ 2 \Sigma_{\varkappa \neq 0} \Sigma_{\lambda \neq 0} \Sigma_{\mu \neq 0} \frac{(H')_{0\varkappa} (\mathbf{p}_i)_{\varkappa\lambda} (H')_{\lambda\mu} (H')_{\mu 0}}{(E_0 - E_\varkappa)(E_0 - E_\lambda)(E_0 - E_\mu)} -$$

$$- 2 \Sigma_{\varkappa \neq 0} \Sigma_{\lambda \neq 0} \frac{(\mathbf{p}_i)_{0\varkappa} H'_{\varkappa 0}}{(E_0 - E_\varkappa)^2} \frac{H'_{0\lambda} H'_{\lambda 0}}{(E_0 - E_\lambda)} - 2 \Sigma_{\varkappa \neq 0} \Sigma_{\lambda \neq 0} \frac{(\mathbf{p}_i)_{0\varkappa} H'_{\varkappa 0}}{(E_0 - E_\varkappa)} \frac{H'_{0\lambda} H'_{\lambda 0}}{(E_0 - E_\lambda)^2}. \quad (28)$$

The various terms are evaluated, using (8) for H', and retaining only terms linear in $\mathbf{E}_0$. All terms on the right-hand side of (28) contain pair contributions, but only the first two involve also triplets. We can write

$$\langle \mathbf{p}_i \rangle^{(3)} = \langle \mathbf{p}_i \rangle^{(3)}_{pairs} + \langle \mathbf{p}_i \rangle^{(3)}_{triplets}.$$

The triplet term can be calculated in much the same way as was done for $\langle \mathbf{p}_i \rangle^{(2)}$. The triplet interactions give, using (23) for the polarizability

$$\langle \mathbf{p}_i \rangle^{(3)}_{triplets} = a_0^3 \Sigma_{k \neq i} \Sigma_{\substack{j \neq k \\ j \neq i}} \mathbf{T}_{ik} \cdot \mathbf{T}_{kj} \cdot \mathbf{E}_0. \quad (29)$$

The triplet term does not affect the polarizability, since, with

$$\boldsymbol{\beta}_i^{(2)} = \boldsymbol{\beta}_{i\,triplets}^{(2)} + \boldsymbol{\beta}_{i\,pairs}^{(2)}$$
$$\boldsymbol{\alpha}_i^{(2)} = \boldsymbol{\alpha}_{i\,triplets}^{(2)} + \boldsymbol{\alpha}_{i\,pairs}^{(2)},$$

and using the relation (20) between $\boldsymbol{\alpha}_i^{(2)}$ and $\boldsymbol{\beta}_i$, one obtains

$$\boldsymbol{\alpha}_{i\,triplets}^{(2)} = 0. \quad (30)$$

The contribution to $\langle \mathbf{p}_i \rangle^{(3)}$ resulting from pairs (i, k) must be evaluated separately for each type of molecule and requires explicit knowledge of the wave functions of the free molecules. The expression for $\langle \mathbf{p}_i \rangle^{(3)}_{pairs}$ is:

$$\langle \mathbf{p}_i \rangle^{(3)}_{pairs} = -2 \Sigma_{k \neq i} \Sigma_{\varkappa \neq 0} \Sigma_{\lambda \neq 0} \Sigma_{\mu \neq 0} \left[\frac{(\mathbf{p}_i)_{0\varkappa} (\mathbf{p}_i \cdot \mathbf{E}_0)_{\varkappa\lambda} (\mathbf{p}_i \cdot \mathbf{T}_{ik} \cdot \mathbf{p}_k)_{\lambda\mu} (\mathbf{p}_i \cdot \mathbf{T}_{ik} \cdot \mathbf{p}_k)_{\mu 0}}{(E_0 - E_\varkappa)(E_0 - E_\lambda)(E_0 - E_\mu)} + \right.$$

$$+ \frac{(\mathbf{p}_i)_{0\varkappa} (\mathbf{p}_i \cdot \mathbf{T}_{ik} \cdot \mathbf{p}_k)_{\varkappa\lambda} (\mathbf{p}_i \cdot \mathbf{E}_0)_{\lambda\mu} (\mathbf{p}_i \cdot \mathbf{T}_{ik} \cdot \mathbf{p}_k)_{\mu 0}}{(E_0 - E_\varkappa)(E_0 - E_\lambda)(E_0 - E_\mu)} +$$

10

$$+ \frac{(\mathbf{p}_i)_{0\kappa} (\mathbf{p}_i \cdot \mathsf{T}_{ik} \cdot \mathbf{p}_k)_{\kappa\lambda} (\mathbf{p}_i \cdot \mathsf{T}_{ik} \cdot \mathbf{p}_k)_{\lambda\mu} (\mathbf{p}_i \cdot \mathbf{E}_0)_{\mu 0}}{(E_0 - E_\kappa)(E_0 - E_\lambda)(E_0 - E_\mu)} +$$

$$+ \frac{(\mathbf{p}_i \cdot \mathbf{E}_0)_{0\kappa} (\mathbf{p}_i)_{\kappa\lambda} (\mathbf{p}_i \cdot \mathsf{T}_{ik} \cdot \mathbf{p}_k)_{\lambda\mu} (\mathbf{p}_i \cdot \mathsf{T}_{ik} \cdot \mathbf{p}_k)_{\mu 0}}{(E_0 - E_\kappa)(E_0 - E_\lambda)(E_0 - E_\mu)} +$$

$$+ \frac{(\mathbf{p}_i \cdot \mathsf{T}_{ik} \cdot \mathbf{p}_k)_{0\kappa} (\mathbf{p}_i)_{\kappa\lambda} (\mathbf{p}_i \cdot \mathbf{E}_0)_{\lambda\mu} (\mathbf{p}_i \cdot \mathsf{T}_{ik} \cdot \mathbf{p}_k)_{\mu 0}}{(E_0 - E_\kappa)(E_0 - E_\lambda)(E_0 - E_\mu)} +$$

$$+ \frac{(\mathbf{p}_i \cdot \mathsf{T}_{ik} \cdot \mathbf{p}_k)_{0\kappa} (\mathbf{p}_i)_{\kappa\lambda} (\mathbf{p}_i \cdot \mathsf{T}_{ik} \cdot \mathbf{p}_k)_{\lambda\mu} (\mathbf{p}_i \cdot \mathbf{E}_0)_{\mu 0}}{(E_0 - E_\kappa)(E_0 - E_\lambda)(E_0 - E_\mu)} \Bigg] +$$

$$+ 2 \Sigma_{k \neq i} \Sigma_{\kappa \neq 0} \Sigma_{\lambda \neq 0} \left[\frac{(\mathbf{p}_i)_{0\kappa} (\mathbf{p}_i \cdot \mathbf{E}_0)_{\kappa 0} (\mathbf{p}_i \cdot \mathsf{T}_{ik} \cdot \mathbf{p}_k)_{0\lambda} (\mathbf{p}_i \cdot \mathsf{T}_{ik} \cdot \mathbf{p}_k)_{\lambda 0}}{(E_0 - E_\kappa)^2 (E_0 - E_\lambda)} + \right.$$

$$\left. + \frac{(\mathbf{p}_i)_{0\kappa} (\mathbf{p}_i \cdot \mathbf{E}_0)_{\kappa 0} (\mathbf{p}_i \cdot \mathsf{T}_{ik} \cdot \mathbf{p}_k)_{0\lambda} (\mathbf{p}_i \cdot \mathsf{T}_{ik} \cdot \mathbf{p}_k)_{\lambda 0}}{(E_0 - E_\kappa)(E_0 - E_\lambda)^2} \right]. \tag{31}$$

In the following sections this expression will be evaluated in a number of special cases.

§ 5. *Coupled harmonic oscillators.* A simple example is the model of an assembly of isotropic harmonic oscillators in an external field $\mathbf{E}_0$, and interacting through induced dipole forces. The wave function for an unperturbed oscillator is given in Cartesian coordinates by the product $u_0(x)u_0(y)u_0(z)$, where $u_0(x) = H_n(\gamma x) e^{-\gamma x^2/2}$, etc., except for a normalizing constant. H_n is the n-th Hermite polynomial, and γ measures the stiffness of the oscillator. The energy difference $E_0 - E_\kappa$ for one oscillator is equal to $- \Sigma nh\nu$, summed over the three coordinates ; $h\nu$ is the spacing of the energy levels and $\Sigma n = \kappa$. With the selection rules

$$(x)_{n,n+1} = \sqrt{\frac{n+1}{2\gamma}} ; \quad (x)_{n,n-1} = \sqrt{\frac{n}{2\gamma}} \tag{32}$$

the summation in (31) can easily be carried out. The result is

$$\langle \mathbf{p}_i \rangle_{pairs}^{(3)} = + a_0^3 \Sigma_{k \neq i} \mathsf{T}_{ik} \cdot \mathsf{T}_{ki} \cdot \mathbf{E}_0, \tag{33}$$

so that we have, from (20), (26) and (33)

$$\alpha_{i\ pairs}^{(2)} = 0. \tag{34}$$

For a system of coupled harmonic oscillators, interacting through induced dipole forces, the polarizability remains constant in second order of approximation.

This result can be generalized to prove that for harmonic oscillators with induced dipole interactions the polarizability remains constant in any order of approximation. (For similar considerations see van Vleck, ref. 7) The total Hamiltonian of N coupled harmonic oscillators in an electric field $\mathbf{E}_0$ is

$$H = \tfrac{1}{2} \Sigma_i a_0^{-1} \mathbf{p}_i^2 + \tfrac{1}{2} \Sigma_i \Sigma_{k \neq i} \mathbf{p}_i \cdot \mathsf{T}_{ik} \cdot \mathbf{p}_k - \Sigma_i \mathbf{p}_i \cdot \mathbf{E}_0, \tag{35}$$

11

where a_0 is the polarizability of a free oscillator. In $3N$-dimensional space this may be written as

$$H = \tfrac{1}{2}a_0^{-1}\mathbf{P}\cdot(\mathbf{U} + a_0\mathbf{T})\cdot\mathbf{P} - \mathbf{P}\cdot\mathbf{E}_0, \tag{36}$$

where $\mathbf{P}$ is the $3N$-dimensional vector in the configuration space of the N oscillators; the projection of this vector on the 3-dimensional subspace of molecule j is $\mathbf{p}_j$. $\mathbf{U}$ is the unit tensor of order $3N \times 3N$. The tensor $\mathbf{T}$ is of order $3N \times 3N$ with elements T_{jk} in 3-dimensional subspace. Conventionally $T_{jk} = 0$ if $j = k$. $\mathbf{E}_0$ is the $3N$-dimensional vector with components $\mathbf{E}_0$ in 3-dimensional subspace. On introducing new vectors $\mathbf{P}'$, defined by

$$\mathbf{P} = \mathbf{S}\cdot\mathbf{P}',$$

where $\mathbf{S}$ is a real orthogonal matrix of order $3N \times 3N$, eq. (36) becomes

$$H = \tfrac{1}{2}a_0^{-1}\mathbf{P}'\cdot\mathbf{S}^{-1}\cdot(\mathbf{U} + a_0\mathbf{T})\cdot\mathbf{S}\cdot\mathbf{P}' - \mathbf{P}'\cdot\mathbf{S}^{-1}\cdot\mathbf{E}_0 \tag{37}$$

By suitable choice of $\mathbf{S}$ the matrix $\mathbf{S}^{-1}\cdot(\mathbf{U} + a_0\mathbf{T})\cdot\mathbf{S}$ can be made diagonal

$$\mathbf{S}^{-1}\cdot(\mathbf{U} + a_0\mathbf{T})\cdot\mathbf{S} = \boldsymbol{\Lambda},$$

where $\boldsymbol{\Lambda}$ is a diagonal matrix of order $3N \times 3N$. The Hamiltonian (37) represents a set of independent harmonic oscillators in an external field $\mathbf{S}^{-1}\cdot\mathbf{E}_0$. The average value of $\mathbf{P}'$ is

$$\langle\mathbf{P}'\rangle = a_0\boldsymbol{\Lambda}^{-1}\cdot\mathbf{S}^{-1}\cdot\mathbf{E}_0$$

and therefore

$$\langle\mathbf{P}\rangle = \mathbf{S}\cdot\langle\mathbf{P}'\rangle = a_0\mathbf{S}\cdot\boldsymbol{\Lambda}^{-1}\cdot\mathbf{S}^{-1}\cdot\mathbf{E}_0 = a_0(\mathbf{U} + a_0\mathbf{T})^{-1}\cdot\mathbf{E}_0,$$

i.e.

$$\langle\mathbf{P}\rangle = a_0(\mathbf{E}_0 - \mathbf{T}\cdot\langle\mathbf{P}\rangle).$$

In 3-dimensional space this equation reads

$$\langle\mathbf{p}_i\rangle = a_0(\mathbf{E}_0 - \Sigma_{k\neq i}\,T_{ik}\cdot\langle\mathbf{p}_k\rangle). \tag{38}$$

The result is that the polarizability of isotropic harmonic oscillators remains unaffected by induced dipole coupling. This property is typical for harmonic oscillators only and cannot be expected to hold for a more realistic molecular model *).

§ 6. *Atomic hydrogen.* The various sums in (31) can, of course, be evaluated in principle if the wave functions of the free atoms or molecules are known with sufficient accuracy. Such a procedure is, however, practically unfeasible even in simple cases. We will use an approximation method which is sufficiently accurate for atomic hydrogen and helium to obtain the correct order of magnitude of the effect of molecular interaction on the polarizability. The

---

*) For an extensive discussion of this point we refer specifically to v a n V l e c k, loc. cit., p. 564—565.

approximation consists in replacing each factor of the denumerators in (31) by a multiple of $-U_0$, where U_0 is an "average excitation energy" of the molecule. If, for instance, the factor $E_0 - E_\varkappa$ refers to the difference in energy between the ground state and the $\varkappa$-th excited state of two molecules, then $E_0 - E_\varkappa$ is replaced by $-2U_0$, with U_0 counted positive. This is a good approximation if all discrete levels of the atom lie within a narrow range on the energy scale (the range should be small compared with U_0). This condition is fulfilled for atomic hydrogen and for helium. To be true, there is a continuum of energy levels which has to be taken into account, but the weight of these levels diminishes rapidly.

With the help of the approximation outlined above, the sums in (31) can be calculated. In Appendix II the various summations are carried out; the result is

$$\langle \mathbf{p}_i \rangle^{(3)}_{pairs} = + \frac{17}{4U_0^3} \Sigma_{k \neq i} [\mathbf{p}_i (\mathbf{p}_i \cdot \mathbf{E}_0) \mathbf{p}_i \cdot \mathsf{T}_{ik} \cdot \mathbf{p}_k \mathbf{p}_i \cdot \mathsf{T}_{ik} \cdot \mathbf{p}_k]_{00} - \tfrac{7}{16} a_0^3 \Sigma_{k \neq i} (\mathsf{T}_{ik} : \mathsf{T}_{ki}) \mathbf{E}_0 +$$

$$+ \tfrac{9}{32} a_0^3 \Sigma_{k \neq i} \mathsf{T}_{ik} \cdot \mathsf{T}_{ki} \cdot \mathbf{E}_0, \qquad (39)$$

where a_0 now stands for

$$a_0 \mathsf{U} = + 2 \frac{(\mathbf{p}_i \mathbf{p}_i)_{00}}{U_0}. \qquad (40)$$

If the same approximation had been used in the evaluation of the second-order approximation in $\langle \mathbf{p}_i \rangle$ and for the third-order due to triplets, the results would have been the same as (25) and (29), but now with a_0 as defined in (40). The first term on the right-hand side of (39) gives (see Appendix III)

$$[\mathbf{p}_i (\mathbf{p}_i \cdot \mathbf{E}_0) \mathbf{p}_i \cdot \mathsf{T}_{ik} \cdot \mathbf{p}_k \mathbf{p}_i \cdot \mathsf{T}_{ik} \cdot \mathbf{p}_k]_{00} = U_0 \frac{a_0}{2} \frac{\{(p_i^a)^4\}_{00}}{3} [2 \mathsf{T}_{ik} \cdot \mathsf{T}_{ik} \cdot \mathbf{E}_0 + (\mathsf{T}_{ik} : \mathsf{T}_{ik}) \mathbf{E}_0], \quad (41)$$

where p_i^a is an arbitrary cartesian component of $\mathbf{p}_i$. In order to rewrite this expression in terms of a_0, we have to use a relation between $\{(p_i^a)^4\}_{00}$ and $\{(p_i^a)^2\}_{00}^2$. For atomic hydrogen this relation is [16]

$$\{(p_i^a)^4\}_{00}/\{(p_i^a)^2\}_{00}^2 = 9/2.$$

Then $\langle \mathbf{p}_i \rangle^{(3)}_{pairs}$ becomes, with (39) and (41)

$$\langle \mathbf{p}_i \rangle^{(3)}_{pairs} = + (15/8) a_0^3 \Sigma_{k \neq i} (\mathsf{T}_{ik} \cdot \mathsf{T}_{ik}) \cdot \mathbf{E}_0 + (23/64) a_0^3 \Sigma_{k \neq i} (\mathsf{T}_{ik} : \mathsf{T}_{ik}) \mathbf{E}_0. \qquad (42)$$

Combining this result with (20) and (26), the change in a_i is

$$\alpha^{(2)}_{i \, pairs} = + (7/8) a_0^3 \Sigma_{k \neq i} \mathsf{T}_{ik} \cdot \mathsf{T}_{ik} + (23/64) a_0^3 \Sigma_{k \neq i} \mathsf{T}_{ik} : \mathsf{T}_{ik} \mathsf{U}. \qquad (43)$$

§ 7. *Helium atoms.* Slater wave functions are used for helium. Noting that here $\mathbf{p}_i = \mathbf{p}_1 + \mathbf{p}_2$, where 1 and 2 refer to the two electrons, the result is

$$\{(p_i^a)^4\}_{00}/\{(p_i^a)^2\}_{00}^2 = 15/4,$$

13

so that

$$\langle \mathbf{p}_i \rangle^{(3)}_{pairs} = +(103/64)\, a_0^3\, \Sigma_{k \neq i}\, (T_{ik} \cdot T_{ik}) \cdot E_0 + (29/128)\, a_0^3\, \Sigma_{k \neq i}\, (T_{ik} : T_{ik})\, E_0, \quad (44)$$

and the change in polarizability in second-order is

$$\boldsymbol{\alpha}^{(2)}_{i\,pairs} = +(39/64)\, a_0^3\, \Sigma_{k \neq i}\, (T_{ik} \cdot T_{ik}) + (29/128)\, a_0^3\, \Sigma_{k \neq i}\, (T_{ik} : T_{ik})\, U, \quad (45)$$

with the same (positive) sign as for hydrogen atoms, although somewhat smaller. We thus find in this order an increase in the polarizability with decreasing intermolecular distances, due to interactions between pairs of molecules. Note that the effect is additive in pairs, i.e. the total effect is equal to the sum of contributions from isolated pairs of atoms or molecules. The next higher order involves clusters of four, three and two molecules, with non-vanishing contributions to the polarizability resulting only from triplets and pairs.

We will show in a subsequent paper that this effect leads to an additional correction to the Clausius-Mosotti function, which is of the same form and of the same order of magnitude, in the density region considered, as the deviations due to statistical fluctuations.

§ 8. *Summary of results and discussion.* The effect of London forces on the polarizability of spherical, nonpolar molecules is obtained from a perturbation expansion in powers $a_0\, T$, where T is the dipole interaction tensor. Since $a_0\, T$ is proportional to a_0/r_{ik}^3 the series converges rapidly for not too high densities. The results obtained for the quantum mechanical average $\langle \mathbf{p}_i \rangle$ of the induced dipole moment in molecule i, are

$$\langle \mathbf{p}_i \rangle^{(1)} = a_0\, E_0,$$

$$\langle \mathbf{p}_i \rangle^{(2)} = -\, a_0^2\, \Sigma_{k \neq i}\, T_{ik} \cdot E_0,$$

$$\langle \mathbf{p}_i \rangle^{(3)} = +\, a_0^3\, \Sigma_{k \neq i}\, \Sigma_{\substack{j \neq k \\ j \neq i}}\, T_{ik} \cdot T_{kj} \cdot E_0 + (C_1 + 1)\, a_0^3\, \Sigma_{k \neq i}\, T_{ik} \cdot T_{ki} \cdot E_0 +$$
$$+ C_2\, a_0^3\, \Sigma_{k \neq i}\, (T_{ik} : T_{ki})\, E_0,$$

where $C_1 = +\ 7/8;\ C_2 = +\ 23/64$ for hydrogen atoms.

$C_1 = +\ 39/64;\ C_2 = +\ 29/128$ for helium atoms.

Collecting terms, the result is up to the third order in $\langle \mathbf{p}_i \rangle$

$$\langle \mathbf{p}_i \rangle = a_0\, [U - a_0\, \Sigma_{k \neq i}\, T_{ik} + a_0^2\, \Sigma_{k \neq i}\, \Sigma_{\substack{j \neq k \\ j \neq i}}\, T_{ik} \cdot T_{kj} +$$
$$+ (C_1 + 1)\, a_0^2\, \Sigma_{k \neq i}\, T_{ik} \cdot T_{ki} + C_2\, a_0^2\, \Sigma_{k \neq i}\, (T_{ik} : T_{ki})\, U] \cdot E_0. \quad (46)$$

We may also write

$$\langle \mathbf{p}_i \rangle = a_0\, [U + C_1\, a_0^2\, \Sigma_{k \neq i}\, T_{ik} \cdot T_{ki} + C_2\, a_0^2\, \Sigma_{k \neq i}(T_{ik} : T_{ki})\, U](E_0 - \Sigma_{k \neq i}\, T_{ik} \cdot \langle \mathbf{p}_k \rangle), \quad (47)$$

where the complete expression for the local field has been used, but the series for the polarizability was broken off after terms in T^2. The results for hydrogen and helium show that the polarizability first increases on the low-density side, with decreasing distances between the molecules. The perturbation expansion of $\langle \mathbf{p}_i \rangle$ is of the same form as the Kirkwood-de Boer series in the statistical theory of the dielectric constant; when a subsequent statistical average is taken it contains additional contributions due to the change in polarizability. The approximation involved in the evaluation of the third-order correction to the induced dipole moment is that all discrete excited energy levels of the unperturbed atom should lie in a range which is small compared with the 'average excitation energy". This condition is fulfilled with sufficient accuracy for atomic hydrogen and helium. The extension of this method to include heavier atoms must rely on a similar approximation method. In addition correlations between the different electrons should be taken into account *). The theory can also be extended to higher multipole interactions, and to optically anisotropic molecules (e.g., diatomic molecules). It may be expected that the results for hydrogen and helium, calculated in this paper, give the general aspects of the effect of second-order forces on the polarizability of nonpolar molecules at least to the correct order of magnitude.

An extension of the Yvon-Kirkwood statistical theory including the change in polarizability with the density in compressed gases, will be presented in a forthcoming paper.

APPENDIX I

On tensor notation

The system of vector and tensor notation used in the present paper is essentially that of M i l n e and C h a p m a n [17]).

Thus the exterior product of an ordered pair of vectors $\mathbf{a}$, $\mathbf{b}$ is a tensor

$$\mathsf{T} = \mathbf{ab} \rightarrow T^{\alpha\beta} = a^\alpha b^\beta$$

The following products of tensors and vectors also occur

$$\mathbf{a} \cdot \mathsf{T} \rightarrow (\mathbf{a} \cdot \mathsf{T})^\alpha = \Sigma_\beta \, a^\beta T^{\beta\alpha}$$
$$\mathsf{T} \cdot \mathbf{a} \rightarrow (\mathsf{T} \cdot \mathbf{a})^\alpha = \Sigma_\beta \, T^{\alpha\beta} a^\beta$$
$$\mathsf{T} \cdot \mathsf{S} \rightarrow (\mathsf{T} \cdot \mathsf{S})^{\alpha\beta} = \Sigma_\gamma \, T^{\alpha\gamma} S^{\gamma\beta}$$
$$\mathsf{T} : \mathsf{S} \Rightarrow \Sigma_{\alpha\beta} \, T^{\alpha\beta} S^{\beta\alpha}$$

In all these formulae the greek superscripts denote the cartesian components of the vectors and tensors.

---

*) If Slater wave functions are used, the following expression holds for spherical, nonpolar molecules

$$\Sigma_{i,j \atop j \neq i} \langle x_i \, x_j \rangle = - \Sigma_{\varkappa,\lambda \atop \lambda \neq \varkappa} |x_{\varkappa\lambda}|^2$$

where i and j label two electrons; $\varkappa$ and λ label the occupied states in the atom, and the wave functions inclusive spins must be used.

Third Order in Dipole Moment

Three of the eight terms occurring in Eq. (31) are evaluated explicitly with the approximation method outlined in the text; the five remaining terms are of the same types as these three. The notation I, II, ... corresponds with the order of terms in (31).

$$I = \Sigma_{\varkappa \neq 0} \Sigma_{\lambda \neq 0} \Sigma_{\mu \neq 0} \frac{(\mathbf{p}_i)_{0\varkappa}(\mathbf{p}_i \cdot \mathbf{E}_0)_{\varkappa\lambda}(\mathbf{p}_i \cdot \mathsf{T}_{ik} \cdot \mathbf{p}_k)_{\lambda\mu}(\mathbf{p}_i \cdot \mathsf{T}_{ik} \cdot \mathbf{p}_k)_{\mu 0}}{(E_0 - E_\varkappa)(E_0 - E_\lambda)(E_0 - E_\mu)} =$$

$$= \Sigma_{\varkappa_i \neq 0} \Sigma_{\lambda_i \neq 0} \Sigma_{\mu_i + \mu_k \neq 0} \frac{(\mathbf{p}_i)_{0\varkappa_i}(\mathbf{p}_i \cdot \mathbf{E}_0)_{\varkappa_i\lambda_i}(\mathbf{p}_i)_{\lambda_i\mu_i} \cdot \mathsf{T}_{ik} \cdot (\mathbf{p}_k)_{0\mu_k}(\mathbf{p}_i)_{\mu_i 0} \cdot \mathsf{T}_{ik} \cdot (\mathbf{p}_k)_{\mu_k 0}}{(E_{0_i} - E_{\varkappa_i})(E_{0_i} - E_{\lambda_i})(E_{0_i} + E_{0_k} - E_{\mu_i} - E_{\mu_k})} \simeq$$

$$\simeq -\frac{1}{2U_0^3} \Sigma_{\varkappa_i \neq 0} \Sigma_{\lambda_i \neq 0} \Sigma_{\mu_i + \mu_k \neq 0} (\mathbf{p}_i)_{0\varkappa_i}(\mathbf{p}_i \cdot \mathbf{E}_0)_{\varkappa_i\lambda_i}(\mathbf{p}_i)_{\lambda_i\mu_i} \cdot \mathsf{T}_{ik} \cdot (\mathbf{p}_k)_{0\mu_k}(\mathbf{p}_i)_{\mu_i 0} \cdot \mathsf{T}_{ik} \cdot (\mathbf{p}_k)_{\mu_k 0}.$$

The summations are performed with matrix algebra. (The state functions with which the matrix elements are formed are a complete set). This gives

$$-\frac{1}{2U_0^3} \Sigma_{\mu_i + \mu_k \neq 0} [\mathbf{p}_i(\mathbf{p}_i \cdot \mathbf{E}_0) \mathbf{p}_i]_{0\mu_i} \cdot \mathsf{T}_{ik} \cdot (\mathbf{p}_k)_{0\mu_k}(\mathbf{p}_i)_{\mu_i 0} \mathsf{T}_{ik} \cdot (\mathbf{p}_k)_{\mu_k 0} +$$

$$+ \frac{1}{2U_0^3} \Sigma_{\mu_i + \mu_k \neq 0} [\mathbf{p}_i (\mathbf{p}_i \cdot \mathbf{E}_0)]_{00} (\mathbf{p}_i)_{0\mu_i} \cdot \mathsf{T}_{ik} \cdot (\mathbf{p}_k)_{0\mu_k}(\mathbf{p}_i)_{\mu_i 0} \cdot \mathsf{T}_{ik} \cdot (\mathbf{p}_k)_{\mu_k 0} =$$

$$= -\frac{1}{2U_0^3} [\mathbf{p}_i (\mathbf{p}_i \cdot \mathbf{E}_0) \mathbf{p}_i \cdot \mathsf{T}_{ik} \cdot \mathbf{p}_k \mathbf{p}_i \cdot \mathsf{T}_{ik} \cdot \mathbf{p}_k]_{00} +$$

$$+ \frac{1}{2U_0^3} [\mathbf{p}_i(\mathbf{p}_i \cdot \mathbf{E}_0]_{00} (\mathbf{p}_i \cdot \mathsf{T}_{ik} \cdot \mathbf{p}_k \mathbf{p}_i \cdot \mathsf{T}_{ik} \cdot \mathbf{p}_k)_{00}.$$

The second term arises because $\lambda_i \neq 0$ in the summation; this term can be evaluated directly, making use of Eq. (40) for the polarizability a_0. The result is

$$I = -\frac{1}{2U_0^3} [\mathbf{p}_i(\mathbf{p}_i \cdot \mathbf{E}_0) \mathbf{p}_i \cdot \mathsf{T}_{ik} \cdot \mathbf{p}_k \mathbf{p}_i \cdot \mathsf{T}_{ik} \cdot \mathbf{p}_k]_{00} + \tfrac{1}{16} a_0^3 (\mathsf{T}_{ik} : \mathsf{T}_{ik}) \mathbf{E}_0.$$

$$II = \Sigma_{\varkappa \neq 0} \Sigma_{\lambda \neq 0} \Sigma_{\mu \neq 0} \frac{(\mathbf{p}_i)_{0\varkappa}(\mathbf{p}_i \cdot \mathsf{T}_{ik} \cdot \mathbf{p}_k)_{\varkappa\lambda}(\mathbf{p}_i \cdot \mathbf{E}_0)_{\lambda\mu}(\mathbf{p}_i \cdot \mathsf{T}_{ik} \cdot \mathbf{p}_k)_{\mu 0}}{(E_0 - E_\varkappa)(E_0 - E_\lambda)(E_0 - E_\mu)} =$$

$$= \Sigma_{\varkappa_i \neq 0} \Sigma_{\substack{\lambda_i + \lambda_k \\ \neq 0}} \Sigma_{\substack{\mu_i + \lambda_k \\ \neq 0}} \frac{(\mathbf{p}_i)_{0\varkappa_i}(\mathbf{p}_i)_{\varkappa_i\lambda_i} \cdot \mathsf{T}_{ik} \cdot (\mathbf{p}_k)_{0\lambda_k}(\mathbf{p}_i \cdot \mathbf{E}_0)_{\lambda_i\mu_i}(\mathbf{p}_i)_{\mu_i 0} \cdot \mathsf{T}_{ik} \cdot (\mathbf{p}_k)_{\lambda_k 0}}{(E_{0i} - E_{\varkappa_i})(E_{0_i} + E_{0_k} - E_{\lambda_i} - E_{\lambda_k})(E_{0_i} + E_{0_k} - E_{\mu_i} - E_{\lambda_k})}.$$

The summation over $\varkappa_i$ is carried out as in I, but for λ_i the term must be split into a part $\lambda_i \neq 0$ and a part $\lambda_i = 0$

$$II \simeq -\frac{1}{4U_0^3} \Sigma_{\lambda_i \neq 0} \Sigma_{\lambda_k \neq 0} \Sigma_{\mu_i + \lambda_k \neq 0} [\mathbf{p}_i \mathbf{p}_i]_{0\lambda_i} \cdot \mathsf{T}_{ik} \cdot (\mathbf{p}_k)_{0\lambda_k}(\mathbf{p}_i \cdot \mathbf{E}_0)_{\lambda_i\mu_i}(\mathbf{p}_i)_{\mu_i 0} \cdot \mathsf{T}_{ik} \cdot (\mathbf{p}_k)_{\lambda_k 0}$$

$$- \frac{1}{2U_0^3} \Sigma_{\lambda_k \neq 0} \Sigma_{\mu_i + \lambda_k \neq 0} [\mathbf{p}_i \mathbf{p}_i]_{00} \cdot \mathsf{T}_{ik} \cdot (\mathbf{p}_k)_{0\lambda_k}(\mathbf{p}_i \cdot \mathbf{E}_0)_{0\mu_i} \cdot (\mathbf{p}_i)_{\mu_i 0} \mathsf{T}_{ik} \cdot (\mathbf{p}_k)_{\lambda_k 0} =$$

$$= - \frac{1}{4U_0^3} [\mathbf{p}_i (\mathbf{p}_i \cdot \mathbf{E}_0) \, \mathbf{p}_i \cdot \mathsf{T}_{ik} \cdot \mathbf{p}_k \mathbf{p}_i \cdot \mathsf{T}_{ik} \cdot \mathbf{p}_k]_{00} -$$

$$- \frac{1}{4U_0^3} [\mathbf{p}_i \, \mathbf{p}_i]_{00} \cdot \mathsf{T}_{ik} \cdot [\mathbf{p}_k (\mathbf{p}_i \cdot \mathbf{E}_0) \, \mathbf{p}_i \cdot \mathsf{T}_{ik} \cdot \mathbf{p}_k]_{00} =$$

$$= - \frac{1}{4U_0^3} [\mathbf{p}_i (\mathbf{p}_i \cdot \mathbf{E}_0) \, \mathbf{p}_i \cdot \mathsf{T}_{ik} \cdot \mathbf{p}_k \mathbf{p}_i \cdot \mathsf{T}_{ik} \cdot \mathbf{p}_k]_{00} - \tfrac{1}{32} a_0^3 \, \mathsf{T}_{ik} \cdot \mathsf{T}_{ik} \cdot \mathbf{E}_0.$$

The remaining terms in (31) give the following results

$$\text{III} = - \frac{1}{2U_0^3} [\mathbf{p}_i(\mathbf{p}_i \cdot \mathbf{E}_0) \, \mathbf{p}_i \cdot \mathsf{T}_{ik} \cdot \mathbf{p}_k \, \mathbf{p}_i \cdot \mathsf{T}_{ik} \cdot \mathbf{p}_k]_{00} - \tfrac{1}{16} a_0^3 \, \mathsf{T}_{ik} \cdot \mathsf{T}_{ik} \cdot \mathbf{E}_0,$$

$$\text{IV} = - \frac{1}{2U_0^3} [\mathbf{p}_i (\mathbf{p}_i \cdot \mathbf{E}_0) \, \mathbf{p}_i \cdot \mathsf{T}_{ik} \cdot \mathbf{p}_k \, \mathbf{p}_i \cdot \mathsf{T}_{ik} \cdot \mathbf{p}_k]_{00} + \tfrac{1}{16} a_0^3 \, (\mathsf{T}_{ik} : \mathsf{T}_{ik}) \, \mathbf{E}_0,$$

$$\text{V} = - \frac{1}{8U_0^3} [\mathbf{p}_i(\mathbf{p}_i \cdot \mathbf{E}_0) \, \mathbf{p}_i \cdot \mathsf{T}_{ik} \cdot \mathbf{p}_k \, \mathbf{p}_i \cdot \mathsf{T}_{ik} \cdot \mathbf{p}_k]_{00} - \tfrac{1}{64} a_0^3 \mathsf{T}_{ik} \cdot \mathsf{T}_{ki} \cdot \mathbf{E}_0,$$

$$\text{VI} = - \frac{1}{4U_0^3} [\mathbf{p}_i(\mathbf{p}_i \cdot \mathbf{E}_0) \, \mathbf{p}_i \cdot \mathsf{T}_{ik} \cdot \mathbf{p}_k \, \mathbf{p}_i \cdot \mathsf{T}_{ik} \cdot \mathbf{p}_k]_{00} - \tfrac{1}{32} a_0^3 \mathsf{T}_{ik} \cdot \mathsf{T}_{ki} \cdot \mathbf{E}_0,$$

$$\text{VII} = \Sigma_{\varkappa \neq 0} \, \Sigma_{\lambda \neq 0} \, \frac{(\mathbf{p}_i)_{0\varkappa} \, (\mathbf{p}_i \cdot \mathbf{E}_0)_{\varkappa 0}}{(E_0 - E_\varkappa)^2} \, \frac{(\mathbf{p}_i \cdot \mathsf{T}_{ik} \cdot \mathbf{p}_k)_{0\lambda} \, (\mathbf{p}_i \cdot \mathsf{T}_{ik} \cdot \mathbf{p}_k)_{\lambda 0}}{(E_0 - E_\lambda)} \simeq$$

$$\simeq - \frac{1}{2U_0^3} \Sigma_{\varkappa_i \neq 0} \, (\mathbf{p}_i)_{0\varkappa_i} (\mathbf{p}_i \cdot \mathbf{E}_0)_{\varkappa_i 0} \Sigma_{\lambda_i \neq 0} \Sigma_{\lambda_k \neq 0} (\mathbf{p}_i)_{0\lambda_i} \cdot \mathsf{T}_{ik} \cdot (\mathbf{p}_k)_{0\lambda_k} (\mathbf{p}_i)_{\lambda_i 0} \cdot \mathsf{T}_{ik} \cdot (\mathbf{p}_k)_{\lambda_k 0} =$$

$$= - \tfrac{1}{16} a_0^3 \, (\mathsf{T}_{ik} : \mathsf{T}_{ki}) \, \mathbf{E}_0,$$

$$\text{VIII} = - \tfrac{1}{32} a_0^3 \, (\mathsf{T}_{ik} : \mathsf{T}_{ki}) \, \mathbf{E}_0.$$

The complete expression (31) is then given by

$$(31) = + \frac{17}{4U_0^3} \Sigma_{k \neq i} [\mathbf{p}_i (\mathbf{p}_i \cdot \mathbf{E}_0) \, \mathbf{p}_i \cdot \mathsf{T}_{ik} \cdot \mathbf{p}_k \mathbf{p}_i \cdot \mathsf{T}_{ik} \cdot \mathbf{p}_k]_{00} +$$

$$+ \tfrac{9}{32} a_0^3 \Sigma_{k \neq i} \mathsf{T}_{ik} \cdot \mathsf{T}_{ki} \cdot \mathbf{E}_0 - \tfrac{7}{16} a_0^3 \Sigma_{k \neq i} (\mathsf{T}_{ik} : \mathsf{T}_{ki}) \, \mathbf{E}_0.$$

APPENDIX III

Proof of Equation (41)

$$[\mathbf{p}_i(\mathbf{p}_i \cdot \mathbf{E}_0) \, \mathbf{p}_i \cdot \mathsf{T}_{ik} \cdot \mathbf{p}_k \mathbf{p}_i \cdot \mathsf{T}_{ik} \cdot \mathbf{p}_k]_{00}^a = \Sigma_{\beta, \gamma, \delta, \varepsilon, \nu} [p_i^\alpha (p_i^\beta E_0^\beta) \, p_i^\gamma \, T_{ik}^{\gamma\nu} \, p_k^\nu \, p_k^\delta \, T_{ik}^{\delta\varepsilon} \, p_i^\varepsilon]_{00}$$

in cartesian coordinates. The summation can be rearranged as follows

$$(p_k^\delta p_k^\delta)_{00} \, \Sigma_{\beta, \gamma, \varepsilon} [p_i^\alpha (p_i^\beta E_0^\beta) \, p_i^\gamma (\mathsf{T}_{ik} \cdot \mathsf{T}_{ik})^{\gamma\varepsilon} \, p_i^\varepsilon]_{00} =$$

$$= a_0 \frac{U_0}{2} \Sigma_{\beta, \gamma, \varepsilon} [p_i^\alpha (p_i^\beta E_0^\beta) \, p_i^\gamma (\mathsf{T}_{ik} \cdot \mathsf{T}_{ik})^{\gamma\varepsilon}) \, p_i^\varepsilon]_{00},$$

since $(p_k^\delta p_k^\delta)_{00}$ is zero for $\nu \neq \delta$ and independent of δ for $\nu = \delta$. The sum vanishes unless α, β, γ, ε are pairwise the same, or are all equal. The following combinations occur

$$\Sigma_{\gamma \neq \alpha} E_0^\alpha [p_i^\alpha p_i^\alpha p_i^\gamma p_i^\gamma]_{00} (\mathsf{T}_{ik} \cdot \mathsf{T}_{ik})^{\gamma\gamma} = [(p_i^\alpha)^2 (p_i^\gamma)^2]_{00} [(\mathsf{T}_{ik} : \mathsf{T}_{ik}) E_0^\alpha - (\mathsf{T}_{ik} \cdot \mathsf{T}_{ik})^{\alpha\alpha} E_0^\alpha], \quad (a)$$

$$\Sigma_{\beta \neq \alpha} E_0^\beta [p_i^\alpha p_i^\beta p_i^\alpha p_i^\beta]_{00} (\mathsf{T}_{ik} \cdot \mathsf{T}_{ik})^{\alpha\beta} = [(p_i^\alpha)^2 (p_i^\beta)^2]_{00} \{[\mathsf{T}_{ik} \cdot \mathsf{T}_{ik} \cdot \mathbf{E}_0]^\alpha - (\mathsf{T}_{ik} \cdot \mathsf{T}_{ik})^{\alpha\alpha} E_0^\alpha\}, \quad (b)$$

$$\Sigma_{\beta \neq \alpha} E_0^\beta [p_i^\alpha p_i^\beta p_i^\beta p_i^\alpha]_{00} (\mathsf{T}_{ik} \cdot \mathsf{T}_{ik})^{\beta\alpha} = [(p_i^\alpha)^2 (p_i^\beta)^2]_{00} \{[\mathsf{T}_{ik} \cdot \mathsf{T}_{ik} \cdot \mathbf{E}_0]^\alpha - (\mathsf{T}_{ik} \cdot \mathsf{T}_{ik})^{\alpha\alpha} E_0^\alpha\}, \quad (c)$$

$$[(p_i^\alpha)^4]_{00} (\mathsf{T}_{ik} \cdot \mathsf{T}_{ik})^{\alpha\alpha} E_0^\alpha. \quad (d)$$

When we make use of $[(p_i^\alpha)^2 (p_i^\beta)^2]_{00} = [(p_i^\alpha)^4]_{00}/3$, the result is

$$[\mathbf{p}_i(\mathbf{p}_i \cdot \mathbf{E}_0) \mathbf{p}_i \cdot \mathsf{T}_{ik} \cdot \mathbf{p}_k \mathbf{p}_i \cdot \mathsf{T}_{ik} \cdot \mathbf{p}_k]_{00} =$$

$$= U_0 \frac{a_0}{2} \tfrac{1}{3} [(p_i^\alpha)^4]_{00} [2\mathsf{T}_{ik} \cdot \mathsf{T}_{ik} \cdot \mathbf{E}_0 + (\mathsf{T}_{ik} : \mathsf{T}_{ik}) \mathbf{E}_0], \quad (41)$$

where p_i^α is an arbitrary cartesian component of $\mathbf{p}_i$.

REFERENCES

1) U h l i g, H. H., K i r k w o o d, J. G. and K e y e s, F. G., J. Chem. Phys. **1** (1933) 155.
2) M i c h e l s, A. and M i c h e l s, C., Phil. Trans. A **231** (1933) 587.
3) M i c h e l s, A. and K l e r e k o p e r, L., Physica **6** (1939) 586.
4) M i c h e l s, A., t e n S e l d a m, C. A. and O v e r d i j k, S. D. J., Physica **17** (1951) 781.
5) M i c h e l s, A. and B o t z e n, A., Physica **15** (1949) 769.
6) L o r e n t z, H. A., "Theory of Electrons" (B. G. Teubner, Leipzig, 1909), p. 137 ff.
7) v a n V l e c k, J. H., J. Chem. Phys. **5** (1937) 320, 556.
8) Y v o n, J., „Recherches sur la Theorie Cinétique des Liquides" (Hermann et Cie., Paris, 1937); for other references to Yvon's work, see Fuller Brown, ref. 11.
9) K i r k w o o d, J. G., J. Chem. Phys. **4** (1936) 592.
10) B o e t t c h e r, C. F., Physica **9** (1942) 945; "Theory of Electric Polarization" (Elsevier Publishing Company, New York, 1952).
11) F u l l e r B r o w n, W., J. Chem. Phys. **18** (1950) 1193, 1200.
12) d e B o e r, J., v a n d e r M a e s e n, F. and t e n S e l d a m, C. A., Physica **19** (1953) 265; t e n S e l d a m, C. A., Thesis, Utrecht (1953).
13) M i c h e l s, A., d e B o e r, J. and B i j l, A., Physica **4** (1937) 981.
14) d e G r o o t, S. R. and t e n S e l d a m, C. A., Physica **13** (1947) 47; **18** (1952) 905, 910.
15) B o r n, M., H e i s e n b e r g, W. and J o r d a n, P., Zeits f. Phys. **35** (1926) 565; see e.g. Condon and Shortley, "The Theory of Atomic Spectra" (Cambridge University Press, 1953) p. 30 ff.
16) Cf. C o n d o n and S h o r t l e y, "The Theory of Atomic Spectra", (Cambridge University Press, 1953).
17) Cf. C h a p m a n, S. and C o w l i n g, T., "The Mathematical Theory of Non-uniform Gases", Cambridge, 1939.

CHAPTER II

ON THE THEORY OF MOLECULAR POLARIZATION IN GASES

II. EFFECT OF MOLECULAR INTERACTIONS ON THE CLAUSIUS-MOSOTTI FUNCTION FOR SYSTEMS OF SPHERICAL NONPOLAR MOLECULES

§ 1. *Introduction.* Accurate measurements of the dielectric constant of compressed gases and liquids have shown that the Clausius-Mosotti relation

$$\frac{\varepsilon - 1}{\varepsilon + 2} V = \tfrac{4}{3} \pi N a \tag{1}$$

fails to hold at high densities. The volume per mole of the system is V and N is Avogadro's number; a is the polarizability of a molecule. The largest deviations occur with polar substances; smaller deviations have been found with nonpolar molecules, to which attention will be restricted in the following calculations. The general tendency for nonpolar molecules is that the Clausius-Mosotti function (1) first increases with increasing density, goes through a maximum at a density of about 200 Amagat and then decreases [1] [2]. The Lorentz-Lorenz expression $(n^2 - 1) V/(n^2 + 2)$ shows the same characteristic behavior [3].

Several theoretical explanations have been offered for these deviations; especially the statistical theory of K i r k w o o d [4] and Y v o n [5] has been successful. In this theory the effect of statistical fluctuations in the induced dipole moment of a molecule is taken into account in successive orders of approximation. Due to this fluctuation effect the Clausius-Mosotti function first increases with increasing density and decreases at higher densities, in qualitative agreement with the experiments. Quantitatively however, the experimental curve increases somewhat more rapidly in the region of low densities and decreases more rapidly at higher densities than the statistical theory predicts. The deviations at high densities can in principle be explained by the effect of first-order interactions between the molecules (repulsive forces) on the polarizability [6] [7]. It was shown in a previous chapter [8] (hereafter referred to as I) that second-order interactions (at-

19

tractive forces) between optically isotropic molecules result in general in an initial increase of the polarizability with decreasing distances between the molecules. An expression for the polarizability was obtained from perturbation theory as a series in powers of the dimensionless quantity $a_0 \mathsf{T}$; T stands for the tensor characteristic for dipole interactions.

$$\mathsf{T}_{ik} = \nabla_i \nabla_k (1/r_{ik}) = r_{ik}^{-3} (\mathsf{U} - (3\mathbf{r}_{ik} \mathbf{r}_{ik}/r_{ik}^2)), \tag{2}$$

where U is the unit tensor, r_{ik} is the distance between the centers of molecules i and k and a_0 is the polarizability of a molecule in the absence of interactions. (The notation in this chapter is the same as in I, unless otherwise specified). In general the polarizability increases in the order T^2, due to interactions between pairs of molecules. (An exceptional case is an assembly of isotropic harmonic oscillators interacting through induced dipole forces; it was shown in I that the polarizability of the oscillators remains constant in any order of approximation). The results obtained in I indicate that better agreement with the experimental data for the Clausius-Mosotti function may be obtained by a combination of the two effects: 1. the change of polarizability with varying distances between the molecules, and 2. the effect of statistical fluctuations, both in the induced dipole moments and in the polarizability of the molecules.

The method we will follow in this chapter is a generalization of a variant of the Yvon-Kirkwood theory proposed by d e B o e r [9]), to include the effect of molecular interactions on the polarizability. The density of the system is supposed to be so low that only the dipole part of the multipole interactions between the isotropic molecules has to be taken into account and that the effect of exchange forces may be neglected.

§ 2. *The Clausius-Mosotti formula for dense gases.* We consider a system of N identical, optically isotropic molecules between the plates of a plane condensor. The electric moment $\mathbf{p}_i$ of the i-th molecule is given by

$$\mathbf{p}_i = a_i \cdot \mathbf{D} - a_i \cdot \Sigma_{k \neq i} \mathsf{T}_{ik} \cdot \mathbf{p}_k, \tag{3}$$

where $\mathbf{D}$ is the electric displacement vector, which for a plane condensor is equal to the external electric field $\mathbf{E}_0$. In this expression $\mathbf{p}_i$ and $\mathbf{p}_k$ represent quantum mechanical averages for a given micro-configuration of the molecules; these averages were denoted by $\langle \mathbf{p}_i \rangle$ and $\langle \mathbf{p}_k \rangle$ in I. The polarizability a_i depends on all intermolecular distances and has been shown in I to be of the form (cf. I, (47))

$$a_i = a_0 [\mathsf{U} + a_0^2 C_1 \Sigma_{k \neq i} \mathsf{T}_{ik} \cdot \mathsf{T}_{ki} + a_0^2 C_2 \Sigma_{k \neq i} (\mathsf{T}_{ik} : \mathsf{T}_{ki}) \mathsf{U}]. \tag{4}$$

Here, a_0 is the polarizability of a free molecule; C_1 and C_2 are constants which are different for different types of molecules. For atomic hydrogen and helium atoms C_1 and C_2 have been evaluated in I. Taking the average of (3), we have

$$\overline{\mathbf{p}}_i = \bar{a}_i \cdot \mathbf{D} - \overline{a_i \cdot \Sigma_{k \neq i} \mathsf{T}_{ik} \cdot \mathbf{p}_k}, \tag{5}$$

which expression may be written as

$$\overline{\mathbf{p}}_i = \bar{a}_i \cdot [\mathbf{D} - \Sigma_{k \neq i} \overline{\mathsf{T}_{ik} \cdot \mathbf{p}_k}] + \bar{a}_i \cdot \Sigma_{k \neq i} \overline{\mathsf{T}}_{ik} \cdot \overline{\mathbf{p}}_k - \overline{a_i \cdot \Sigma_{k \neq i} \mathsf{T}_{ik} \cdot \mathbf{p}_k}. \qquad (6)$$

For the average value of the induced dipole moment $\mathbf{p}_i$ we have $\overline{\mathbf{p}}_i = \overline{\mathbf{p}}_k = \overline{\mathbf{p}}$ and similarly for the average polarizability $\bar{a}_i = \bar{a}_k = \bar{a}\mathsf{U}$, where $\bar{a}$ is the scalar average polarizability. Further it can be shown that [4])

$$\Sigma_{k \neq i} \overline{\mathsf{T}}_{ik} \cdot \overline{\mathbf{p}}_k = (8\pi/3)\, n\overline{\mathbf{p}} = (8\pi/3)\, \mathbf{P}, \qquad (7)$$

where $n = N/V$ is the number of molecules per unit of volume and $\mathbf{P}$ is the molecular polarization. The last two terms on the right hand side of (6) represent the effect of statistical fluctuations. Whereas in the Yvon-Kirkwood theory there are only fluctuations in the dipole moments, the expression (6) now includes also contributions from the fluctuations in the polarizability of a molecule. These fluctuations arise because the mean moment or polarizability of a molecule known to be at a specified position with respect to another molecule is not the same as the corresponding quantities of that molecule without such specification (for an extensive discussion of this point see e.g. Fuller Brown, ref. 5). If all fluctuations could be neglected (6) gives directly the Clausius-Mosotti relation (1), but now with the *average* polarizability instead of the polarizability a_0 of a free molecule. It will be shown that one may write

$$\bar{a}\, \Sigma_{k \neq i} \overline{\mathsf{T}}_{ik} \cdot \overline{\mathbf{p}}_k - \overline{a_i \cdot \Sigma_{k \neq i} \mathsf{T}_{ik} \cdot \mathbf{p}_k} = \tfrac{1}{3}\, \bar{a}\, (\varepsilon + 2)\, \mathbf{E}R, \qquad (8)$$

where $\mathbf{E}$ is the electric field strength and R is a function $R(n, T)$ of density and temperature only. For constant polarizability R reduces to the function $S(n, T)$ introduced by d e B o e r [9]), which takes into account fluctuations in the dipole moments only.

Multiplying (6) by n one obtains, with (7) and (8)

$$(\varepsilon - 1)\, \mathbf{E}/4\pi = \tfrac{1}{3}(\varepsilon + 2)\, \mathbf{E}\, n\bar{a} + \tfrac{1}{3}(\varepsilon + 2)\, \mathbf{E}R\, n\bar{a}, \qquad (9)$$

since $\mathbf{P} = (\varepsilon - 1)\, \mathbf{E}/4\pi$ and $\mathbf{D} - 8\pi\mathbf{P}/3 = \tfrac{1}{3}(\varepsilon + 2)\, \mathbf{E}$.

We then obtain the following formal expression for the Clausius-Mosotti function

$$\frac{\varepsilon - 1}{\varepsilon + 2}\, V = \tfrac{4}{3}\pi N\bar{a}[1 + R(n, T)]. \qquad (10)$$

This equation shows that the Clausius-Mosotti function $(\varepsilon - 1)V/(\varepsilon + 2)$ changes with density for two reasons: 1. the average polarizability is a function of density; 2. statistical fluctuations occur, both in the induced dipole moments and in the polarizabilities of the molecules. As follows from (4) the average polarizability can be written as a power series in a_0. We may therefore use the following alternative form for the right hand side of (10)

$$\frac{\varepsilon - 1}{\varepsilon + 2}\, V = \tfrac{4}{3}\pi N a_0 \,[1 + R'(n, T)], \qquad (11)$$

where $R'(n, T)$ contains the total functional dependence of the Clausius-Mosotti expression on density and temperature. It will be shown below that also $R'(n, T)$ can be written as a series in powers of a_0.

In the statistical theory with constant polarizability, a_0, the expression obtained by de Boer for the Clausius-Mosotti function is

$$\frac{\varepsilon - 1}{\varepsilon + 1} V = \tfrac{4}{3} \pi N a_0 [1 + S(n, T)], \tag{12}$$

where S is the correction arising from fluctuations in the dipole moments only. Note that the difference between (11) and (12) is not caused only by the change of the average polarizability with density, but contains also fluctuation terms in the polarizabilities of the molecules.

However, it will appear that the first nonvanishing term in $R'(n, T)$ does not yet contain fluctuations in the polarizability, but only a contribution from the relative change in the average polarizability and a term arising from fluctuations in the induced dipole moments. This justifies the procedure followed by t e n   S e l d a m   and   d e   G r o o t [10] who calculate the change of the average polarizability with density by comparing the theoretical curve computed from (12) with the experimental data for argon. In the next section we will first calculate the average polarizability as a function of density for helium atoms.

§ 3. *The average polarizability as a function of density.* Up to the order T^2 the expression for the average polarizability is, from (4)

$$\bar{a} \mathsf{U} = a_0 \left[\mathsf{U} + a_0^2 C_1 \Sigma_{k \neq i} \overline{\mathsf{T}_{ik} \cdot \mathsf{T}_{ki}} + a_0^2 C_2 \Sigma_{k \neq i} \overline{(\mathsf{T}_{ik} : \mathsf{T}_{ki})} \, \mathsf{U} \right]. \tag{13}$$

The following statistical expressions can be given for the averages in (13)

$$\Sigma_{k \neq i} \overline{\mathsf{T}_{ik} \cdot \mathsf{T}_{ki}} = n \int g(r_{ik}) \, \mathsf{T}_{ik} \cdot \mathsf{T}_{ki} \, d\mathbf{r}_k, \tag{14}$$

$$\Sigma_{k \neq i} \overline{\mathsf{T}_{ik} : \mathsf{T}_{ki}} = n \int g(r_{ik}) \, \mathsf{T}_{ik} : \mathsf{T}_{ki} \, d\mathbf{r}_k. \tag{15}$$

Here $g(r_{ik})$ is the well-known distribution function for pairs of molecules. Eq. (14) may be rewritten as

$$\Sigma_{k \neq i} \overline{\mathsf{T}_{ik} \cdot \mathsf{T}_{ki}} = \mathsf{U} \frac{n}{3} \int g(r_{ik}) \, \mathsf{T}_{ik} : \mathsf{T}_{ki} \, d\mathbf{r}_k. \tag{16}$$

We then have for the relative change in the average polarizability, with (13), (15) and (16)

$$a_0^2 C_1 \Sigma_{k \neq i} \overline{\mathsf{T}_{ik} \cdot \mathsf{T}_{ki}} + a_0^2 C_2 \Sigma_{k \neq i} \overline{\mathsf{T}_{ik} : \mathsf{T}_{ki}} \, \mathsf{U} =$$
$$= a_0^2 \, \mathsf{U} \, (C_1/3 + C_2) \, n \int g(r_{ik}) \, \mathsf{T}_{ik} : \mathsf{T}_{ki} \, d\mathbf{r}_k. \tag{17}$$

The expansion for $g(r_{ik})$ in term of the molecular density n is

$$g(r_{ik}) = e^{-\beta \varphi_{ik}} (1 + n g'(r_{ik}) + O(n^2)),$$

with
$$g'(r_{ik}) = \int (e^{-\beta \varphi_{il}} - 1) (e^{-\beta \varphi_{kl}} - 1) \, d\mathbf{r}_l. \tag{18}$$

Here $\beta = 1/kT$ and φ_{ik} represents the intermolecular potential between two molecules. Expanding the relative change in the average polarizability in powers of the density

$$\overline{\Delta a}/a_0 = (\bar{a} - a_0)/a_0 = a(T)n + b(T)n^2, \tag{19}$$

We obtain for the coefficients $a(T)$ and $b(T)$

$$a(T) = a_0^2 \, (C_1/3 + C_2) \int e^{-\beta\varphi_{ik}} \, \mathbf{T}_{ik} : \mathbf{T}_{ki} \, \mathrm{d}\mathbf{r}_k =$$
$$= 8\pi a_0^2 \, (C_1 + 3C_2) \int_0^\infty e^{-\beta\varphi_{ik}} \, r_{ik}^{-4} \, \mathrm{d}r_{ik}, \tag{20}$$
$$b(T) = 8\pi a_0^2 \, (C_1 + 3\,C_2) \int_0^\infty e^{-\beta\varphi_{ik}} \, g'(r_{ik}) \, r_{ik}^{-4} \, \mathrm{d}r_{ik}. \tag{21}$$

The integrals occurring in $a(T)$ and $b(T)$ have been evaluated by d e B o e r, v a n d e r M a e s e n and t e n S e l d a m (loc. cit.) for a Lennard Jones potential field

$$\varphi(r) = 4\varepsilon \, [(\sigma/r)^{12} - (\sigma/r)^6],$$

(ε is the depth of the potential well at the minimum energy and $\varphi(\sigma) = 0$) as a function of the "reduced" temperature $T^* = kT/\varepsilon$. The values of the

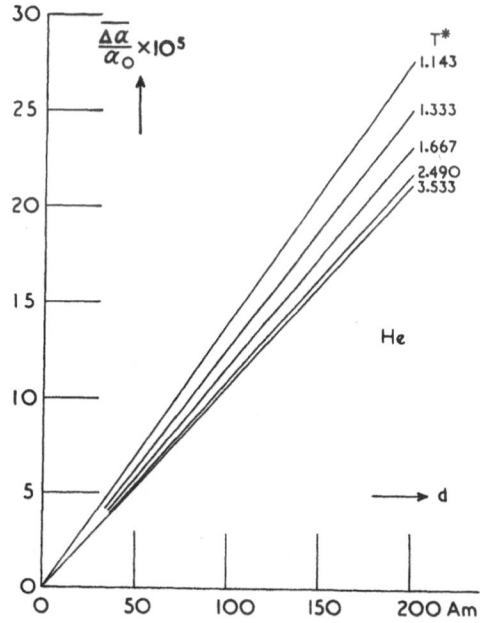

Fig. 1. Relative change of the average polarizability of helium atoms with density, times 10^5, for five "reduced" temperatures.

parameters ε and σ are known for a large number of gases from experiments e.g. on the second virial coefficients. The constants C_1 and C_2 have been determined in I for atomic hydrogen and helium atoms; since atomic hydrogen is an unrealistic example we will restrict ourselves in this section to helium atoms.

With the help of experimental values of a_0, ε and σ we calculate the relative change in the average polarizability for helium atoms at low temperatures. For convenience the values of the parameter T^* will be used for which d e B o e r c.s. evaluated the integrals (20) and (21). Since the theory presented in I does not hold for very high densities (contributions from clusters of more than three molecules are neglected) we will use only the values 50, 100, 150 and 200 Amagat units for the density. The results are given in the graph below; $\varepsilon/k = 10.22$ degrees for helium.

The relative increase in the polarizability of helium atoms is seen to be of the order of one hundredth of one percent up to densities of 200 Amagat. For heavier atoms, e.g. argon, the increase is much larger, since $a(T)$ and $b(T)$ are proportional to a_0^2 and the polarizability of argon atoms is about eight times as large as that of helium. Therefore if C_1 and C_2 are of the same order of magnitude for argon was for helium, as may be expected, the relative increase in the polarizability of argon atoms is of the order of one tenth of one percent in the same density region.

This increase in the average polarizability with density, due to London forces between the molecules, gives rise to a correction term in the Clausius-Mosotti function (1) which is of the same order of magnitude as the effect due to fluctuations in the dipole moments. In the next section we will discuss the Clausius-Mosotti function (10) or its alternative form (11).

§ 4. *Evaluation of the correction term $R(n, T)$.* The function $R(n, T)$ on the right hand side of Eq. (8) can be evaluated by expanding the left hand side of that equation in successive orders of approximation. For that purpose we define an auxiliary quantity $\mathbf{p}_i'$ by

$$\mathbf{p}_i' = \boldsymbol{\alpha}_i \cdot [\mathbf{D} - \Sigma_{k \neq i} \overline{\mathsf{T}}_{ik} \cdot \overline{\mathbf{p}_k}] = \tfrac{1}{3}(\varepsilon + 2) \, \boldsymbol{\alpha}_i \cdot \mathbf{E}, \qquad (22)$$

i.e. $\mathbf{p}_i'$ is the (quantum mechanical) average dipole moment which *would* be induced in molecule i *if* the other molecules had during their motion always the average moment $\overline{\mathbf{p}}_i = \overline{\mathbf{p}}$. By combining (3) and (22) we have

$$\mathbf{p}_i = \mathbf{p}_i' - \boldsymbol{\alpha}_i \cdot (\Sigma_{k \neq i} \mathsf{T}_{ik} \cdot \mathbf{p}_k - \Sigma_{k \neq i} \overline{\mathsf{T}}_{ik} \cdot \overline{\mathbf{p}_k}). \qquad (23)$$

When this expression for $\mathbf{p}_i$ is inserted back into the left hand side of (8) we obtain

$$\bar{a}\, \Sigma_{k \neq i} \overline{\mathsf{T}}_{ik} \cdot \overline{\mathbf{p}_k} - \overline{\boldsymbol{\alpha}_i \cdot \Sigma_{k \neq i} \mathsf{T}_{ik} \cdot \mathbf{p}_k} =$$

$$= \Sigma_{k \neq i} (\bar{a}\, \overline{\mathsf{T}}_{ik} \cdot \overline{\mathbf{p}_k'} - \overline{\boldsymbol{\alpha}_i \cdot \mathsf{T}_{ik} \cdot \mathbf{p}_k'}) + \Sigma_{k \neq i} \overline{\boldsymbol{\alpha}_i \cdot \mathsf{T}_{ik} \cdot \boldsymbol{\alpha}_k \cdot \Sigma_{l \neq k} \mathsf{T}_{kl} \cdot \mathbf{p}_l} - \qquad (24)$$

$$- \Sigma_{k \neq i} \bar{a}\, \overline{\mathsf{T}}_{ik} \cdot \overline{\boldsymbol{\alpha}_k \cdot \Sigma_{l \neq k} \mathsf{T}_{kl} \cdot \mathbf{p}_l} - \Sigma_{k \neq i} (\overline{\boldsymbol{\alpha}_i \cdot \mathsf{T}_{ik} \cdot \boldsymbol{\alpha}_k} - \bar{\boldsymbol{\alpha}}_i \cdot \overline{\mathsf{T}}_{ik} \cdot \bar{\boldsymbol{\alpha}}_k) \cdot \Sigma_{l \neq k} \overline{\mathsf{T}}_{kl} \cdot \overline{\mathbf{p}_l}.$$

The next step in this approximation procedure is to replace each $\mathbf{p}$ by $\mathbf{p}'$ and to add a fluctuation term of higher order, with the help of (23). At any stage of the process one may replace each $\mathbf{p}$ by $\mathbf{p}'$ and break the series off, thus making only a very small error. Then it is seen from (22) that the left hand side of (8) is proportional to $(\varepsilon + 2)\,\mathbf{E}$, and thus also the right hand side.

Next we substitute for a_i in (24) the expression (4) derived in I and evaluate $\bar{a}R$ up to terms in T^2. The result is

$$\bar{a}R(n,\,T)\mathsf{U} = a_0^3\,\Sigma_{k\neq i}\,\Sigma_{l\neq k}\,(\overline{\mathsf{T}_{ik}\cdot\mathsf{T}_{kl}} - \overline{\mathsf{T}_{ik}}\cdot\overline{\mathsf{T}_{kl}}), \tag{25}$$

and the formula for the Clausius-Mosotti function becomes

$$\frac{\varepsilon - 1}{\varepsilon + 2}\,V\mathsf{U} = \tfrac{4}{3}\pi N a_0\,[\mathsf{U} + a_0^2\,C_1\,\Sigma_{k\neq i}\,\overline{\mathsf{T}_{ik}\cdot\mathsf{T}_{ki}} + a_0^2\,C_2\,\Sigma_{k\neq i}\,\overline{\mathsf{T}_{ik}:\mathsf{T}_{ki}}\mathsf{U} + \tag{26}$$

$$+ a_0^2\,\Sigma_{k\neq i}\,\Sigma_{l\neq k}\,(\overline{\mathsf{T}_{ik}\cdot\mathsf{T}_{kl}} - \overline{\mathsf{T}_{ik}}\cdot\overline{\mathsf{T}_{kl}})].$$

The correction to the Clausius-Mosotti function (1) consists of two parts in this order of approximation. One part is due to fluctuations in the induced dipole moments and the second part is caused by the change in the average polarizability of the molecules by interaction. Fluctuations in the polarizability enter the series expansion only in higher orders. As was mentioned before, this justifies the semi-empirical procedure followed by d e G r o o t and t e n S e l d a m who calculated the change in the average polarizability by comparing values of the Clausius-Mosotti function with the theoretical expression (12), but with a_0 replaced by $\bar{a}$. In higher orders this procedure is not valid, as is seen directly from (24).

The right hand side of (25) is the same as de Boer's function $a_0 S_2(n,\,T)$. For the evaluation of (26) we add the results of the calculations on S_2 to the change in the average polarizability with density as determined in the previous section for helium atoms. Following de Boer the expression for S_2 is written as a power series in the density

$$S_2(n,\,T) = a_2(T)n + b_2(T)n^2 + \ldots\ldots \tag{27}$$

The coefficients a_2 and b_2 are closely related to the coefficients a and b which occur in the series for the average polarizability. In addition, however, b_2 contains contributions from triplets of molecules. For the total dependence of the Clausius-Mosotti function on the density we can write, in this order of approximation

$$R'(n,\,T) = S_2(n,\,T) + \overline{\varDelta a}/a_0. \tag{28}$$

In the following table the function $R'(n,\,T)$ is listed for helium atoms, at densities of 50, 100, 150 and 200 Amagat units and for two reduced temperatures $T^* = 1.33$ and $T^* = 2.49$. The values of S_2 were taken from the paper by d e B o e r (loc. cit.)and the values of $\overline{\varDelta a}/a_0$ are those plotted in fig. 1.

It is seen from the table that the correction to the Clausius-Mosotti function due to the change in the average polarizability of the molecules with the density is of the same order of magnitude as the effect of fluctuations in the induced dipole moments. The effect increases with increasing density and decreases somewhat with increasing temperature, as should be expected.

TABLE I §)

$\overline{\varDelta a}/a_0$, S_2 and R', times 10^4, for helium atoms as a function of density and of the reduced temperature T^*						
d_{Amagat}	T^* 1.333			T^* 2.490		
	$\overline{\varDelta a}/a_0$	S_2	R'	$\overline{\varDelta a}/a_0$	S_2	R'
50	0.63	0.47	1.10	0.55	0.41	0.95
100	1.26	0.90	2.17	1.09	0.79	1.88
150	1.90	1.29	3.19	1.63	1.14	2.78
200	2.53	1.64	4.17	2.18	1.47	3.65

§) The values of n are those for an ideal gas, which involves only a small error for helium.

§ 5. *The Clausius-Mosotti function for argon.* Although the coefficients C_1 and C_2 in the series expansion for the polarizability were determined in I only for atomic hydrogen and for helium atoms, their values may be expected to be of the same order of magnitude for argon. In this section we compare the experimental results for the Clausius-Mosotti and Lorentz-Lorenz *) functions for argon with the theory presented in this paper. D e B o e r c.s. (loc. cit.) have compared the Yvon-Kirkwood statistical theory with the experimental results for argon over a wide range of densities and for two reduced temperatures 1.333 and 2.490 (corresponding with 160 and 298°K respectively, for argon). In the following graph the Yvon-Kirkwood theory is compared with the experimental data by M i c h e l s, t e n S e l d a m and O v e r d ij k and M i c h e l s and B o t z e n (ref. 3). The deviations between theory and experiments at high densities ($>$ 200 Am) can in principle be explained by the effect of first-order interactions on the polarizability of the molecules (see refs. 6 and 7). To explain the deviations at low densities (up to 200 Amagat) we calculate the average change in the polarizability, leaving the values of C_1 and C_2 open. The following table gives the results for $\varDelta a/a_0(C_1 + 3C_2)$ at 50, 100, 150 and 200 Amagat and for $T^* = 1.333$ and $T^* = 2.49$.

From a comparison between the values in the table and the deviations between the statistical theory and the experimental values for the Clausius-Mosotti and Lorentz-Lorenz functions we deduce that $C_1 + 3C_2 \simeq 1$, which is of the same order of magnitude as for helium ($C_1 + 3C_2 = 165/128$). It may therefore be expected that the theory presented in this paper will improve

*) The Lorentz-Lorenz expression may also be taken for the comparison between the statistical theory and the experimental results because the effect of dispersion is probably very small.

the agreement between theory and experiments for the Clausius-Mosotti function of compressed argon.

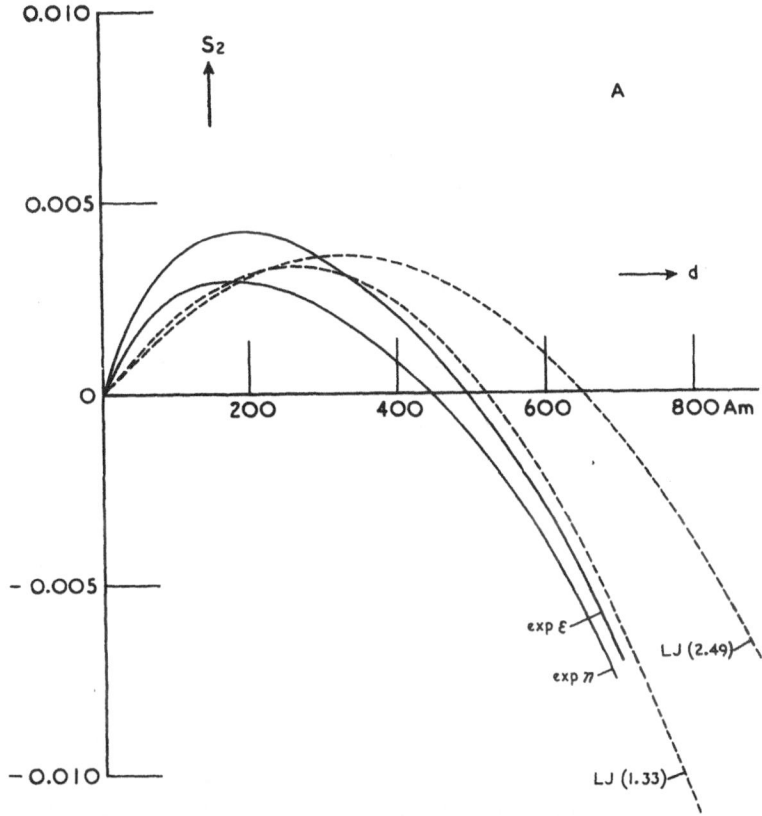

Fig. 2. Comparison of the density dependence of the Clausius-Mosotti and Lorentz-Lorenz functions, as obtained from argon data on the dielectric constant ε and the index of refraction n, with the theoretical values calculated from the Yvon-Kirkwood statistical theory by d e B o e r c.s. [9]) for the "reduced" temperatures $T^* = 1.333$ and 2.490 (160°K and 298°K, respectively, for argon).

TABLE II *)

$\Delta a/a_0 (C_1 + 3C_2)$, times 10^3, for argon as a function of density and reduced temperature				
$d_{A\,magat}$	50	100	150	200
$T^* = 1.333$	1.267	2.529	3.787	5.040
$T^* = 2.490$	1.092	2.184	3.275	4.365

*) For argon $a_0 = 16.39 \times 10^{-25}$ cm³; $\sigma = 3.405$ Å, $\sigma^3 n_0 = 0.0010619$, where n_0 is the number of molecules per cm³ at N.T.P. (see C. A. ten Seldam, Thesis, Utrecht, 1953).

§ 6. *Summary and conclusions.* A calculation of the Clausius-Mosotti function for compressed, optically isotropic gases was presented, taking into account the change in the polarizability of the molecules by interactions at low densities. It was found that the Clausius-Mosotti function increases at low densities with increasing density. This increase is larger than calculated on the basis of the Yvon-Kirkwood statistical theory with density independent polarizability. For helium atoms the effect of molecular interactions on the polarizability gives a correction to the Clausius-Mosotti function which is of the same order of magnitude as the effect due to fluctuations in the induced dipole moments. For argon atoms no accurate theoretical values are available, but it may be expected that the theory leads to an essential improvement compared with the statistical theory for density independent polarizability. The analysis of the Clausius-Mosotti function was split into two parts:

I. a quantum mechanical calculation of the induced dipole moment in successive orders of approximation from perturbation theory;

II. a subsequent classical average over a canonical ensemble.

The applicability of the Yvon-Kirkwood statistical theory and of the modified version presented here is restricted to low densities (since contributions from clusters of more than three molecules were neglected). At higher densities the formalism is mathematically too complex and the convergence of the series may be expected to be slow. In addition it would not be sufficiently accurate to restrict the interaction operator between the molecules to the induced dipole term; the perturbation should include higher induced multipoles.

For these reasons an extension of the theory to higher densities along these lines is hardly promising at the outset. On the other hand, the high density method of d e G r o o t and t e n S e l d a m [7]) is again based on a consideration of *individual molecules*; it takes into account molecular interactions as an effect on the average polarizability of a molecule (assuming that it is possible to use an ordinary potential term in the Hamiltonian instead of an antisymmetric zero-order wave function), and it furthermore takes over the low density statistical fluctuations in the induced depole moments, neglecting fluctuations in the polarizability of the molecules.

From a more fundamental point of view it would be more satisfactory if instead of an analysis based on individual molecules (which implies necessarily a two-step procedure), the theory could be based on a (quantum) statistical treatment on an assembly of electrons and nuclei.

REFERENCES

1) Uhlig, H. H., Kirkwood, J. G. and Keyes, F. G., J. Chem. Phys. **1**, (1933) 155.
2) Michels, A. and Michels, C., Phil. Trans. A **231** (1933) 587; Michels, A. and Kleerekoper, L., Physica **6** (1939) 586.
3) Michels, A., ten Seldam, C. A. and Overdijk, S. D. J., Physica **17** (1951) 781; Michels, A. and Botzen, A., Physica **15** (1949) 769.
4) Kirkwood, J. G., J. Chem. Phys. **4** (1936) 592.
5) Yvon, J., "Recherches sur la Theorie Cinétique des Liquides" (Hermann et Cie., Paris, 1937); Fuller Brown, W., J. Chem. Phys. **18** (1950) 1193, 1200.
6) Michels, A., de Boer, J. and Bijl, A., Physica **4** (1937) 981.
7) de Groot, S. R. and ten Seldam, C. A., Physica **13** (1947) 47; **18** (1952) 905, 910; ten Seldam, C. A., Thesis, Utrecht (1953).
8) Chapter I of this thesis.
9) de Boer, J., van der Maesen, F. and ten Seldam, C. A., Physica **19** (1933) 265.
10) ten Seldam, C. A. and de Groot, S. R., Physica **18** (1952) 910; ten Seldam, C. A., Thesis, Utrecht, 1953.

Chapter III

DEVIATIONS FROM ADDITIVITY OF THE INTERMOLECULAR FIELD AT HIGH DENSITIES

§ 1. *Introduction*. For the statistical evaluation of physical properties of compressed gases, of liquids and solids the intermolecular field *) is usually assumed to be additive, i.e. the interaction field may at all densities be written as a sum of terms referring to isolated pairs of molecules. This statement is known as the "additivity of intermolecular forces". The assumption of additivity is obviously not valid for molecules which tend to associate and for molecules which form hydrogen bonds. We restrict ourselves to forces between spherically symmetric atoms and we exclude from consideration the types of forces mentioned above.

Since the smallest number of molecules for which nonadditive contributions may arise is three, we consider a group of three identical spherically symmetric atoms or molecules, a, b and c, with closed shells of electrons. The interaction between the atoms will be evaluated by applying perturbation theory, on the basis of the valence bond method, which is useful for simple systems. The zero-order wave function is written as

$$\Psi_0 = \Sigma_\lambda (- 1)^\lambda P_\lambda \varphi_a \varphi_b \varphi_c, \tag{1}$$

except for a normalizing constant. The wave function φ_a is a solution of the wave equation

$$E_{0a} = \int \varphi_a^* H_{0a} \varphi_a \, \mathrm{d}\tau,$$

where E_{0a} is the energy eigenvalue for the ground state of a free atom a, and H_{0a} is the Hamilton operator for the free atom. Further, P_λ is any permutation operator of the symmetric group on $3n$ particles; n is the number of electrons per molecule. The group consists of $(3n)!$ elements and λ is even or odd for even and odd permutations, respectively. The operator P_λ can be written as

$$P_\lambda = P_{\lambda a} P_{\lambda b} P_{\lambda c} P_{\lambda abc},$$

---

*) The term "intermolecular field" refers to the interaction energy of a special microconfiguration of molecules and not to an ensemble average. Even if the assumption of additivity is valid then the ensemble averages can in general not be written in terms of contributions from isolated pairs of molecules.

where $P_{\lambda a}$ refers to a permutation between electrons of atom a, etc., and $P_{\lambda abc}$ is a permutation between electrons on different atoms. The Hamiltonian of the system of three interacting atoms is

$$H = H_0 + H'_{abc}, \tag{2}$$

where H_0 is the Hamilton operator for three noninteracting atoms and H'_{abc} is the interaction operator. First-order perturbation theory gives the repulsive interaction between the atoms. From the calculations by P. R o - s e n [1]) it appears that first-order forces between three helium atoms are not equal to the sum of interactions between isolated pairs; this means that repulsive forces between spherical atoms do not have the property of additivity *). The larger the "overlap" of wave functions the greater is the nonadditive contribution. Therefore the nonadditivity is larger in the case of an equilateral triangle of atoms than for other configurations having the same triangular perimeter. R o s e n obtained the following results

$$\frac{E_{abc}}{E_{ab} + E_{ac} + E_{bc}} = -1.15 \; e^{-0.33(3R)}$$

for the equilateral configuration; R is the distance between pairs in units of Bohr radii. The nonadditive term is E_{abc}; E_{ab} is the first-order interaction for an isolated pair ab, etc. For the linear symmetric case the result is

$$\frac{E_{abc}}{E_{ab} + E_{ac} + E_{bc}} = +9.8 \; e^{-0.66(4R)},$$

with R, $2R$ the distances between pairs. For the equilateral triangle the error involved in neglecting nonadditivity is one percent or less if $R \geqslant 4.8$ Bohr radii. The error involved for the linear symmetric configuration has the opposite sign and is always less than for the equilateral triangle in the region considered ($R \geqslant 1.8$ Bohr radii). For large separations $E_{abc}/3E_{ab} = -I$ is a good approximation in the equilateral triangular case; I is the overlap integral. Therefore the effect of nonadditivity is negligible if the overlap integral is small compared with unity.

On increasing the distances between the atoms the first-order forces decrease rapidly. The second-order interactions become predominant, and the evaluation of these forces may at large distances be based on a zero-order wave function which is a simple product of atomic wave functions

$$\Psi_0 = \Psi_a \, \Psi_b \, \Psi_c, \tag{3}$$

with

$$\Psi_a = \Sigma_{\lambda a} \, (-1)^{\lambda a} \, P_{\lambda a} \, \varphi_a,$$

and the Hamiltonian

$$H = H_0 + H'_{abc}.$$

---

*) This could have been stated a priori, since first-order forces are the roots of secular-equations which are in general irrational.

L o n d o n [2]) and M a r g e n a u [3]) have shown that second-order forces are additive, if the zero-order wave function may be written as in (3), i.e. if overlap of wave functions may be neglected. A x i l r o d and T e l l e r [4]) [5]) applied third-order perturbation theory to the interactions between neutral atoms, again for a zero-order wave function (3). It was observed that this order reflects the interactions between triplets of molecules, giving rise to a nonadditive effect in the dipole dispersion forces. The magnitude of this so-called triple dipole interaction can be illustrated by comparing it with the dipole interaction for three rare gas atoms, as calculated on the basis of three isolated pairs. For convenience we consider again two configurations: an equilateral triangle and a collinear array. The results are

$$E_{abc}/(E_{ab} + E_{ac} + E_{bc}) = - (11/32)\, \alpha/R^3$$

for the equilateral triangle, and $+ (4/43)\, \alpha/R^3$ for the collinear array. The second-order forces between isolated pairs are E_{ab}, E_{ac} and E_{bc} and are counted negative. The polarizability of a free atom is α; R and $2R$ are again the distances between pairs. The results show that the attractive field is decreased in the equilateral triangle and increased for the linear symmetric array, compared with an additive sum over isolated pairs. For the crystals of neon, argon, krypton and xenon, $(11/32)\, \alpha/R^3$ is equal to 0.0041, 0.0099, 0.0139 and 0.0166, respectively. A x i l r o d [6]) summed the third-order interaction energy for crystals of the heavy rare gases. This energy is positive, thus decreasing the attraction between the molecules, and amounts to two to nine percent of the cohesive energy of the crystals of the heavy rare gases. It is not possible on this basis to explain why the rare gases (except helium) crystallize in the face centered cubic structure instead of in the slightly more dense lattice of hexagonal symmetry. A x i l r o d evaluated the difference in triple dipole energy between the two lattice types and found that, although this energy favors the cubic structure, the difference is only of the order of one tenthousandth of the cohesive energy and hence does not explain the absolute stability of the face centered cubic lattice (see also the next chapter).

The nonadditivity calculations of R o s e n and A x i l r o d are straight-forward extensions of the evaluation of first- and second-order interactions between two atoms or molecules. Two remarks may be made in connection with the third-order interactions. First, at the densities where the nonadditive effect is not negligible, the interaction may not be restricted to induced dipoles only, but must include poles of higher order. Second, it is doubtful whether a zero-order wave function of the form (3) is still a good starting point at the densities of the crystals, where the repulsive forces are of the order of one half of the attractive field. This indicates that a zero-order wave function should be used which is at least antisymmetric with respect to nearest neighbors. For instance, if of the three atoms a, b, c two are close together (a, b) and one is far apart (c), then the zero-order wave function should be written

$$\Psi_0 = (\Sigma_{\lambda ab} (- 1)^{\lambda ab} P_{\lambda ab} \Psi_a \Psi_b) \Psi_c, \tag{5}$$

instead of (3), i.e. the second-order forces between (ab) and (c), as calculated on the basis of isolated pairs, may be different from the result obtained with a wave function of the form (5). The purpose of the following calculation is to indicate a method by which the effect of exchange terms on the second-order interactions may be estimated, based on the model of the "caged" atom or molecule.

§ 2. *Formalism.* We consider a system of N identical, spherically symmetric atoms $a, b, \ldots, N$ with closed shells of electrons. The model of the caged atom is based on the assumption that the exchange interactions between electrons on different atoms may be replaced by an ordinary potential term (see also discussion at the end of this chapter). For the evaluation of second-order forces between neutral molecules this implies that we may start the perturbation calculation from a zero-order wave function which is a simple product of atomic wave functions

$$\Psi_d = \Psi_{da} \Psi_{db} \Psi_{dc} \ldots . \tag{6}$$

and the Hamiltonian

$$\mathsf{H} = \mathsf{H}_0 + H', \tag{7}$$

where H' is the usual interaction operator. The Hamiltonian H_0 includes the potential term for the exchange interaction between electrons on different atoms, with

$$\mathsf{H}_0 = \Sigma_a \mathsf{H}_{0a},$$

and where

$$\mathsf{H}_{0a} = H_{0a} + V_a(r). \tag{9}$$

The Hamiltonian for a free atom a is H_{0a}, and $V_a(r)$ is the potential term replacing the effect of exchange; r is the distance from the center of atom a. The wave functions Ψ_{da} are solutions of the equations

$$H_{0a} \Psi_{da} = (E_{0a} + E_a') \Psi_{da}. \tag{10}$$

The energy of the ground state of a free atom is E_0 and E_a' is the first-order interaction between atom a and its neighbors; E_a' is a function of the density of the system. Solutions of the wave equations (10) are possible if e.g. $V_a(r)$ has the form of a boundary condition

$$\begin{aligned} V_a(r) &= 0 \quad \text{for} \quad r < R \\ &= \infty \;\text{for} \quad r \geqslant R; \end{aligned} \tag{11}$$

R is called the "radius of the cage" of atom a. In principle, the value of R can be determined if E_{0a} and E'_a are known (from solutions for the free atom and of the determinantal wave equation for the first-order interaction). In practice one solves (10), with the condition (11), for different values of R. Then it is assumed that the pressure

$$P = - (1/4\pi R^2)\partial E'_a/\partial R,$$

which the electron "gas" exerts on the wall of the cage, counterbalances the external pressure of the whole gas[7]). In doing so, a rough model of the effect of exchange terms is used. The wave equations (10) have been solved for hydrogen atoms [8-10]), helium atoms [11]), the hydrogen molecule ion [12]) and argon atoms [13]).

§ 3. *Second-order forces*. With the new zero-order wave functions Ψ_{da} perturbation theory is applied up to the second order. The first-order change in energy is identically zero, and in second order the dispersion forces are again additive, *but now with respect to the new unperturbed state of caged atoms*. If we write for the attractive field between a pair of caged atoms $\varphi_{ij}^{(d)}$, then we have for the total second-order interactions

$$\varphi^{(d)} = \tfrac{1}{2} \Sigma_{i,j}\, \varphi_{ij}^{(d)}. \tag{12}$$

On the other hand, the dispersion forces in a system consisting of isolated pairs of atoms are

$$\varphi^{(0)} = \tfrac{1}{2} \Sigma_{i,j}\, \varphi_{ij}^{(0)}. \tag{13}$$

If $\varphi^{(d)} \neq \varphi^{(0)}$, then the dispersion forces are nonadditive with respect to isolated pairs. London's expression for the dipole forces between two atoms is

$$\varphi_{ij}^{(0)} = - 3\, V_0\, \alpha_0^2/4r_{ij}^6, \tag{14}$$

where V_0 is the first ionization potential of a free atom and α_0 is its polarizability; r_{ij} is the distance between the centers of the two atoms. In the same approximation we obtain for the dipole forces between a pair of "caged" atoms

$$\varphi_{ij}^{(d)} = - 3\, V_d\, \alpha_d^2/4r_{ij}^6. \tag{15}$$

Here, V_d is the first ionization potential of a caged atom in its unperturbed state and α_d is the polarizability of a caged atom. As follows from the calculations on the caged atom or molecule, both V_d and α_d are smaller than their values for the free atoms. Physically this means that exchange effects compress the electron clouds and this results in a decrease of the intrinsic dipole moment and in an increase of the fundamental frequencies of the electrons.

Consequently, the second-order forces are weaker than calculated for isolated pairs. If we take a pair of atoms ij, then a measure of the nonadditive effect will be given by

$$(\varphi_{ij}^{(d)} - \varphi_{ij}^{(0)})/\varphi_{ij}^{(0)} = 2\Delta a/a_0 + \Delta V/V_0, \tag{16}$$

where we have substituted $a_d = a_0 + \Delta a$, $V_d = V_0 + \Delta V$, and only linear terms have been taken into account. The values of $\Delta a/a_0$ and of $\Delta V/V_0$ must be taken from the theoretical calculations on the caged atom; for argon from references 7 and 13. At a temperature of 25°C and a density of 600 Amagat $\Delta a/a_0 = -3.2 \times 10^{-2}$. It is difficult to give an good estimate for $\Delta V/V_0$, but for helium we may replace $\Delta V/V_0$ by approximately $\Delta E/E_0$, where E_0 is the energy of the ground state of a free helium atom. Reference 11 then gives for helium at a pressure of 5×10^4 atm $\Delta E/E_0 = -0.014$ and $\Delta a/a_0 = -0.168$. For argon we will assume also that $\Delta a/a_0$ is much larger than $\Delta V/V_0$ and we will take as a lower limit of $2\Delta a/a_0 + \Delta V/V_0$ a value -6×10^{-2}. This means that at 25°C and a density of 600 Amagat the dipole dispersion forces in compressed argon are of the order of six percent weaker than calculated on the assumption of additivity.

Equation (16) may be rewritten in terms of any set of intermolecular parameters; for instance we have for a Lennard Jones potential field $\varphi_{ij}^{(0)} = -4\varepsilon(\sigma/r_{ij})^6$, and a corresponding expression for $\varphi_{ij}^{(d)}$. The change in the quantity $\varepsilon\sigma^6$ is then

$$\Delta(\varepsilon\sigma^6) = (2\Delta a/a_0 + \Delta V/V_0)\,(\varepsilon\sigma^6)^{(0)} \tag{17}$$

where the superscript (0) refers to the values of the intermolecular parameters as determined from low density measurements of the second virial coefficients, etc.

§ 4. *Thermodynamic functions.* For the calculation of thermodynamic functions we write for the second-order interactions between two molecules, including the nonadditive effect in the dipole dispersion forces.

$$\varphi_{ij}^{(d)} = \varphi_{ij}^{(0)} + (2\Delta a/a_0 + \Delta V/V_0)\,(-4\varepsilon\sigma^6/r_{ij}^6), \tag{18}$$

with values of ε and σ as determined from low density data. Next it is assumed that the atoms may be considered fixed at the centers of their cells as far as the nonadditive part of (18) is concerned, i.e. the distance r_{ij} may be replaced by the distance between the centers of cells i and j, R_{ij}. With this approximation the partition function is a product of two parts and the thermodynamic functions are additive in the two parts of the field. It should be noted that the nonadditive effect is independent of temperature in this approximation. For the internal energy $U = U_0 + \Delta U$ we obtain

$$\Delta U = -2N(2\Delta a/a_0 + \Delta V/V_0) \times (\varepsilon\sigma^6/R_0^6) \cdot \Sigma_{j=2}^{N}(R_0/R_{1j})^6, \tag{19}$$

where R_0 is the distance between nearest neighbors in the lattice. The value of the lattice sum in (19) is 14.4 for a face centered cubic array [14]; further $R_0 = 4.3$ Å at a density of 600 Amagat. The values ε and σ, as determined from low density pvT data, are 165×10^{-16} ergs and 3.405 Å, respectively [15]). When these values are inserted into (19) we obtain for ΔU about $+100$ cal per mole.

On account of the model used for the calculation of this nonadditive effect this value is probably too high. The result indicates, however, that the use of a simple-product type of wave function for the evaluation of second-order forces at high densities overestimates the magnitude of the dipole dispersion interactions.

§ 5. *Relative magnitude of effects*. In terms of perturbation theory the nonadditivity in the repulsive forces is a first-order effect, the effect of exchange terms on the dispersion forces between atoms is of the second order, whereas Axilrod's nonadditive effect is of third order. Evidently this does not imply that the relative magnitude of these effects has always this same order. The nonadditivity in the first-order forces will predominate at the highest densities. In the crystals of the heavy rare gases the first-order interactions are only about half as large as the forces of second order and the three nonadditive effects may be of comparable magnitude. Quantitative comparison between these calculations and experimental values of the sublimation energies of crystals at absolute zero is possible only if the inter-molecular field between *two* molecules is accurately known. It will not be attempted here to give such a comparison on the basis of the Lennard Jones potential field, because there are indications that this form of the interaction field does not accurately represent the molecular interaction at small distances between the molecules (see further chapter V).

Another aspect of the possible importance of nonadditive contributions to the intermolecular field at high densities may be mentioned here; it will be discussed in more detail in the next chapter. K i h a r a [16]) has shown that a Lennard Jones (6, n) potential field fails to explain the absolute stability of the fcc structure as compared with the hexagonal crystal of closest packing, for any acceptable value of n, if additivity of intermolecular forces is assumed. For an additive "exp-six" potential of the form

$$\varphi(r) = \frac{\varepsilon}{1 - 6/a} \left[\frac{6}{a} e^{a(1 - r/r_0)} - \left(\frac{r_0}{r} \right)^6 \right], \tag{20}$$

however, the cubic structure can become more stable than the hexagonal lattice if $a < 8.675$ (K i h a r a, loc. cit. *). The separation between two molecules is r; ε is the depth of the potential well, r_0 is the position of the minimum and a is a parameter which measures the steepness of the repulsion energy. To avoid effects due to nonadditivity at high densities, the accurate determination of the parameters ε, a and r_0 must be based on experimental second virial coefficients and measurements of transport properties at moderate densities and at high temperatures. This has been done recently by

---

*) K i h a r a assumes that the zero-point energy does not play any essential role regarding the stability of the face centered cubic structure since helium, despite its large zero-point energy, crystallizes in the hexagonal closest packing.

M a s o n and R i c e [11] for helium and hydrogen. They found that a potential of the form (20) is definitely superior to a Lennard Jones potential. A combination of high temperature measurements at moderate densities with accurate data on the sublimation energies of the crystals at $0°K$ and on zero-point energies will give information on the importance of nonadditive effects at high densities.

§ 6. *Discussion*. The results of the foregoing analysis indicate that deviations from additivity of the intermolecular field are not negligible at the densities of the crystals of the heavy rare gases; the same result must be expected with diatomic or polyatomic molecules. However, considerable care must be exercised in the interpretation of these results. The analysis was based on the method of the caged atom or molecule; in this method it is assumed that the effect of exchange between electrons on different molecules may be replaced by an ordinary potential term. The validity of this procedure lacks theoretical proof. A comparison between experimental results for compressed argon and calculations of the change in kinetic energy of the electrons and of the polarizability as a function of density, based on the caged atom model, has been made especially by t e n S e l d a m and d e G r o o t [7] [13]). It appears that the caged atom model predicts correctly the qualitative behaviour of these physical properties at high densities. Quantitative information about the validity of the assumption inherent in the caged atom model is, however, much more difficult to obtain. First, the caged atom method is based on a *static* model, in which the effect of density fluctuations is neglected. It was discussed in Chapter II that fluctuations in the local density of the system play an important role for the explanation of the change in the Clausius-Mosotti function with density and that the density dependence of the *average* polarizability of the molecules can be determined in a simple way only in the density region where contributions from clusters of more than three molecules may be neglected. Second, the Yvon-Kirkwood-de Boer statistical theory, supplemented to include the effect of molecular interactions (Chapter II), is essentially a low-density calculated and can, therefore, not be used at the high densities where the caged atom model is usually applied. Fortunately the effect of fluctuations is probably not important at high densities (beyond 400 Amagat), but for the second reason mentioned the Clausius-Mosotti function is not a suitable physical property for the comparison with calculations based on the caged atom model.

An interesting suggestion has recently been made by N ij b o e r [13]). Instead of measurements on the electric susceptibility of compressed gases, he proposes that diamagnetic susceptibilities be determined. In view of the smallness of the induced magnetic moments, collective effects (fluctuation phenomena) may be neglected and therefore the internal field may be taken equal to the external field. It follows then that the diamagnetic susceptibility per unit mass is given by

$$\chi_d = - (Ne^2/6mc^2M) \Sigma_{i=1}^n \overline{r_i^2},$$ (21)

where M is the molecular weight, N is Avagadro's number, n is the number of electrons per molecule and $\overline{r_i^2}$ is the averaged square of the radial distance of the i-th electron to the nucleus. Comparison of this expression with the Kirkwood variational expression for the polarizability of a molecule

$$a = (4/9na_0) (\Sigma_{i=1}^n \overline{r_i^2})^2$$ (22)

(a_0 is the radius of the first Bohr orbit of hydrogen) shows that the polarizability is directly proportional to the square of the diamagnetic susceptibility.

It would then be possible to separate the contributions due to intermolecular (collective) from those due to intra-molecular effects ("individual" effects, i.e. the influence of interactions on the physical properties of an "individual" atom or molecule) to the deviations from constancy observed for the Clausius-Mosotti expression as a function of pressure. As a result the predictions of a "collective" theory such as Kirkwood's, as well as those of an "individual" theory like the caged atom model of d e G r o o t and t e n S e l d a m could be tested separately.

This prediction is, however, based on the *additional* assumption that the influence of intermolecular forces figures in the Clausius-Mosotti function only as an *average* change of the polarizability of a molecule, i.e. that for the evaluation of the effect of molecular interactions each molecule may be treated individually. It has been shown in Chapter II of this thesis that this assumption is only valid in the region of densities where contributions from clusters of more than three molecules may be neglected. In general the effect of molecular interactions on the Clausius-Mosotti function contains "collective" terms, due to *fluctuations in the interaction field* between the molecules.

More *direct* information can be obtained from measurements of the change in kinetic energy of the electrons with the density. This information is taken from pvT-data by an application of the "virial theorem", which is based on a statistical consideration of an *assembly of nuclei and electrons*. Difficulties arising from the "individuality of the molecules" do not occur and the virial theorem is therefore valid both at low at high densities. From a more fundamental point of view it would be more satisfactory to reformulate the theory of molecular polarization in terms of electrons and nuclei (see also end of chapter II).

The results of the analysis of this chapter may be summarized as follows: It can be proven that the intermolecular field at high densities deviates in general from the assumption of additivity. An approximate evaluation of these deviations for second-order forces between neutral molecules can be given in a general way *if it may be assumed* that the exchange forces between electrons on different atoms may be replaced by an ordinary potential term. (Even if this assumption is valid, then the caged atom model gives only approximate results because of the special form chosen for the exchange potential function).

It must be expected that better knowledge about the importance of non-additive effects (i.e. "manybody" or "multibody" interactions) in the inter-molecular field can be obtained from a consideration of the stability of crystal structures for the rare gases; this will be attempted in the next chapter. A general treatment of this problem of stability of crystal structures was under-taken by B o r n and co-workers [21]. Instead of considering the problem which structures are stable and which are unstable, we will discuss the more special question why the heavy rare gases crystallize in a face centered cubic lattice and not in the almost identical structure of hexagonal closest packing. This problem has to do with the *precise* form of the intermolecular field and with the importance of multibody interactions *).

REFERENCES

1) R o s e n, P., J. Chem. Phys. **21** (1953) 1007.
2) L o n d o n, F., Z. Physik. Chem. **B 11** (1930) 222.
3) M a r g e n a u, H., Revs. Modern Phys. **11** (1939) 1.
4) A x i l r o d, B. M. and Teller, E., J. Chem. Phys. **11** (1943) 299.
5) A x i l r o d, B. M., J.Chem. Phys. **17** (1949) 1349); **19** (1951) 719.
6) A x i l r o d, B. M., J. Chem. Phys. **19** (1951) 724.
7) T e n S e l d a m, C. A., Thesis, Utrecht (1953).
8) M i c h e l s, A., d e B o e r, J. and B ij l, A., Physica **4** (1937) 981.
9) d e G r o o t, S. R. and t e n S e l d a m, C. A., Physica **12** (1946) 669.
10) M i c h e l s, A. and d e G r o o t, S. R., Physica **16** (1950) 183.
11) d e G r o o t, S. R. and t e n S e l d a m, C. A., ref. 9.
12) S o m m e r f e l d A. and W e l k e r, H., Ann. Physik **32** (1938) 56.
13) t e n S e l d a m, C. A. and d e G r o o t, S. R., Physica **18** (1952) 891, 905.
14) C o t t r e l l, T. L., Trans. Faraday Soc. **47** (1951) 337.
15) t e n S e l d a m, C. A. and d e G r o o t, S. R., Physica **18** (1952) 910.
16) L e n n a r d J o n e s, J. E. and I n g h a m, A. E., Proc. Roy. Soc. (London) A **107**(1925) 636.
17) M i c h e l s, A. and W ij k e r, H., Physica **15** (1949) 627.
18) K i h a r a, T., Revs. Modern Phys. **25** (1953) 831.
19) M a s o n, E. A. and R i c e, W. E., J. Chem. Phys. **22** (1954) 522.
20) N ij b o e r, B. R. A., Physica **19** (1953) 454, see also d e G r o o t, S. R. Ned. Tijdschr. Nat. **19** (1953) 259.
21) B o r n c.s., M., Series of Papers in Proc. Phil. Soc. Cambridge (1940–1942), "On the Stability of Crystal Lattices".

---

*) A direct evaluation of second-order forces for a group of three atoms, two of which overlap appreciably, seems to be of interest for nonadditivity considerations. A calculation for helium atoms has been undertaken; preliminary results show that induced dipole forces between one atom and the two overlapping atoms are not the same as calculated for two isolated pairs and may deviate con-siderably from additivity.

Chapter IV

ON INTERMOLECULAR FORCES
AND CRYSTAL STRUCTURES OF THE RARE GASES

§ 1. *Introduction.* For the accurate determination of intermolecular parameters from experimental data, two models for the potential field have been
used most frequently. The first model is the Lennard Jones $(s, 6)$ potential

$$\varphi(r) = \frac{\varepsilon}{s - 6} [6(r_0/r)^s - s(r_0/r)^6].\tag{1}$$

The distance between the centers of two molecules is r and ε is the depth of
the potential minimum (at $r = r_0$). The second model is the "modified"
Buckingham potential $(a, 6)$, also called the "exp-six" potential, which may
be written in the form

$$\varphi(r) = \frac{\varepsilon}{a - 6} [6e^{a(1 - r/r_0)} - a(r_0/r)^6],\tag{2}$$

where a is a parameter which measures the steepness of the repulsion
energy. The second potential is somewhat more realistic than the first, since
theoretical calculations give an exponential decrease of the first-order interactions with increasing r. C o r n e r [1]), employing a Buckingham potential
with an additional term for the induced dipole-quadrupole interaction (varying
as r^{-8}) has shown how the various parameters may be determined accurately
for neon and argon by combining gas properties and crystal data. In the
calculation of the crystal energies, C o r n e r took into account the influence
of zero-point energy. M a s o n and R i c e [2]) extended this method to
include experimental transport properties for a number of simple, nonpolar
molecules, using the modified Buckingham potential as a model. The resulting
values of a are: neon 14.5; argon 14.0; krypton 12.3 and xenon 13.0. These
values are of interest for the following calculations.

The question which we will discuss in this chapter is whether a potential
function of the form (1) or (2) can explain the observed crystal structures of
the rare gases. As is known from experiments, neon, argon, krypton and xenon
crystallize in a face centered cubic lattice, whereas helium forms hexagonal
crystals under pressure.

L e n n a r d J o n e s and I n g h a m [3]) calculated the potential energy of molecular crystals for three types of cubic lattices: simple cubic, cubic body centered and cubic close packed (face centered cubic). They found that for a potential of the form

$$\varphi(r) = \lambda_n/r^n - \lambda_m/r^m; \quad m < n$$

the face centered cubic structure is more stable than the body centered, and a body centered more stable than a simple cubic lattice. However, there are two structures of closest packing, cubic and hexagonal. The number of nearest neighbors in both structures is 12; the number of next nearest neighbors, which are $2^{\frac{1}{2}}$ times farther away from the central molecule, is 6, both in the cubic and in the hexagonal lattice. In the cubic crystal the molecules in the third shell are $3^{\frac{1}{2}}$ times farther away from the central molecule, but in the hexagonal lattice this distance is only $(2 + \frac{2}{3})^{\frac{1}{2}}$ times the distance between nearest neighbors.

K i h a r a and K o b a [4]) extended the calculations by comparing the potential energies of the hexagonal and the cubic close packings. These authors used the potential functions (1) and (2) with the parameter s varying between 7 and 18, and a in (2) between 8.5 and 18. Their results for the Lennard Jones potential show that in this case the hexagonal lattice is more stable than the cubic structure for any value of s. The results for the Buckingham potential are particularly interesting, since they show that for $a < 8.765$ the face centered cubic lattice is more stable than the hexagonal packing, whereas for $a > 8.765$ the hexagonal crystal is the stable one.

Two remarks should be made in connection with these results. First, the limiting value of a is much lower than the experimental values for the rare gases neon, argon, krypton and xenon, which crystallize in the face centered cubic structure. Second, the value $a \simeq 8.765$ is very extreme for the modified Buckingham potential, since the lowest value of a which gives a potential minimum is about 8.5. In addition, Kihara and Koba neglected the effect of zero-point energy on the stability of the crystal structures because helium, despite its large zero-point energy, crystallizes in the hexagonal lattice (see K i h a r a, ref. 5). As a further simplification the intermolecular field was assumed to be additive (i.e. the interaction energy may be calculated as a sum over isolated pairs of molecules).

As possible explanations for the discrepancy between the theoretical and experimental values of a it should be mentioned first that the inclusion of zero-point energy may shift the limiting value of a appreciably. Second, it is known from theoretical considerations that the intermolecular field is not strictly additive [5]). Since manybody forces are not spherically symmetric, their contribution to the crystal energies of the two lattice types may be different for the following reasons:

1) F o r r e a s o n s o f s y m m e t r y . Directional differences may

already be manifest in the arrangement of the first neighbors. An explanation in this direction has been suggested by P r i n s [6]). He supposes that a compression of the 8-electron shell of the heavy rare gases is in the cubic structure more favorable than in the crystal of hexagonal symmetry.

2) F o r r e a s o n s o f d e n s i t y. Nonadditive contributions weaken the attractive forces between the molecules. The two first shells of molecules are at the same distances from the central molecule in the two lattices, but the third shell is somewhat closer to the central molecule in the hexagonal lattice. Therefore, nonadditive contributions to the attractive forces tend to stabilize the cubic lattice. A x i l r o d [7]) evaluated the third-order non-additive effect for the noble gases and compared its magnitude for the fcc and hcp crystals. It was found that this effect favors the fcc structure, but the difference between the two lattices is only of the order of one tenthousandth of the cohesive energy and hence cannot explain the absolute stability of the fcc structure.

Accurate calculations of other nonadditive effects are not available. We therefore restrict ourselves mainly to the determination of the total energy of the crystals at absolute zero, with the inclusion of zero point energy, assuming that the assumption of additivity is valid and that the intermolecular forces have spherical symmetry.

§ 2. *Determination of zero-point energy.* The zero-point energy of a molecular crystal is calculated using the method proposed by C o r n e r [8]). In the model Corner uses, the molecules carry out harmonic oscillations about their equilibrium positions. The frequency of oscillation is calculated for one molecule with the other molecules at rest. The following expression for the zero-point energy is then obtained

$$U_{zp}/\varepsilon = F(s) \, \Lambda^* \left[s(s-1) \left(\frac{r_0}{R_0} \right)^{s+2} C_{s+2} - 5s \left(\frac{r_0}{R_0} \right)^{8} C_8 \right]^{\frac{1}{2}} \qquad (3)$$

for a Lennard Jones potential $(s, 6)$, with

$$F(s) = \frac{9}{8} \left(\frac{5}{3} \right)^{\frac{1}{2}} \Big/ \left\{ [2\pi^2(s-6)]^{\frac{1}{2}} \left(\frac{s}{6} \right)^{1/(s-6)} \right\} \qquad (4)$$

$$\Lambda^* = h/\sigma \, (m\varepsilon)^{\frac{1}{2}}. \qquad (5)$$

The intermolecular distance between two molecules for zero potential is σ and m is the mass of a molecule. Further, R_0 is the nearest neighbor distance in the crystal, and the C_s are crystal sums for the fcc lattice, tabulated by L e n n a r d J o n e s and I n g h a m [3]) and by K i h a r a and K o b a [4]). In case of a hexagonal lattice, analogous equations hold for the zero-point energy, with C_s replaced by the hexagonal crystals sums H_s. Values for the crystal sums H_s were determined by G o e p p e r t-M a y e r and K a n e [9]) and K i h a r a and K o b a.

The range $\Lambda^* = 0.1$ to 0.7 covers all rare gases except helium. The zero-point energy of helium is so large that it cannot be calculated on the basis of Corner's model of harmonic oscillations; we will therefore not discuss the helium crystal. For the Buckingham potential (2) the expression for the zero-point energy is

$$U_{zp}/\varepsilon = G(a)\, \Lambda^* \left(\frac{a}{a-6}\right)\left[\left(\frac{a}{2} - \frac{1}{R^*}\right) e^{a(1-R^*)} - 5\frac{C_8}{R^{*8}}\right]^{\frac{1}{2}}, \qquad (6)$$

where

$$G(a) = \frac{9}{8}\left(\frac{5}{3}\right)^{\frac{1}{4}} \Big/ \left[2\pi^2\left(\frac{a}{a-6}\right)\right]^{\frac{1}{4}} x(a), \qquad (7)$$

and $x(a) = r_0/\sigma$ is a solution of the transcendental equation

$$6 \exp\left(a - a/x\right) = a\, x^6; \qquad (8)$$

R^* is the reduced lattice distance R_0/r_0. Values for $G(a)$ can be computed from a table given by Hirschfelder and Rice [10] for the range $a = 12.0$ to 15.0. These values, completed with those for $a = 10.0$ and $a = 16.0$, are listed in the following table I.

TABLE I

Reduced separation σ/r_0 at which the modified Buckingham potential is zero, and values for the function $G(a)$		
a	$1/x = \sigma/r_0$	$G(a)$
10.0	.8547 4895	.1767 1798
12.0	.8761 0051	.2025 1215
14.0	.8891 0396	.2197 0790
16.0	.8986 0719	.2322 3200

We have not taken into consideration values for a lower than 10.0, since in this region the modified Buckingham potential becomes unrealistic. Moreover, the experimental values for a lie all above 12.0. For the hcp lattice the expression for the zero-point energy is the same as (6), except that C_8 is replaced by H_8.

§ 3. *The crystal energies at absolute zero.* The expression for the zero-point energy must be added to the potential energy of the crystals, to give the total or sublimation energy. As was stated before we assume that the molecular forces are additive. For a Lennard Jones potential the energy per molecule is

$$U_{pot}/\varepsilon = \frac{1}{2(s-6)}\left[6\left(\frac{r_0}{R_0}\right)^s C_s - s\left(\frac{r_0}{R_0}\right)^6 C_6\right] \qquad (9)$$

for a fcc lattice. In case of a hcp structure the quantities C_s, C_6 are replaced by H_s and H_6. The distance R_0 between nearest neighbors is determined by adding (3) and (9) and differentiating with respect to r_0/R_0, i.e.

$$\partial(U_{pot} + U_{zp})/\partial\left(\frac{r_0}{R_0}\right) = 0$$

or

$$0 = \frac{6s}{s-6}\left[\left(\frac{r_0}{R_0}\right)^{s-1}C_s - \left(\frac{r_0}{R_0}\right)^5 C_6\right] +$$

$$+ F(s)\,\varLambda^* \frac{\left[s(s-1)\,(s+2)\left(\frac{r_0}{R_0}\right)^{s+1}C_{s+2} - 40s\left(\frac{r_0}{R_0}\right)^7 C_8\right]}{\left[s(s-1)\left(\frac{r_0}{R_0}\right)^{s+2}C_{s+2} - 5s\left(\frac{r_0}{R_0}\right)^8 C_8\right]^{\frac12}}. \quad (10)$$

This equation is solved numerically by trial and error. An analogous expression holds for the hcp lattice, again with C_s replaced by H_s, etc. The equilibrium value of r_0/R_0 is then substituted into the sum of (3) and (9) to give the total energy of the crystal at absolute zero.

For the Buckingham potential the expression for the potential energy per molecule is

$$U_{pot}/\varepsilon = \frac{1}{2(a-6)}\left[72\,\varTheta\,e^{a(1-R^*)} - a\,\frac{C_6}{R^{*6}}\right], \quad (11)$$

where $\varTheta$ is a function of aR^*, introduced by C o r n e r [1] and defined by

$$\Sigma_i\,e^{a(1-r_i/r_0)} = 12\varTheta\,e^{a(1-R^*)}.$$

The summation extends over all the lattice sites of the crystal. Corner tabulated values of $\varTheta$ for the range $aR^* = 10.5$ to 16 and a fcc crystal. His results have been supplemented with those for $aR^* = 8, 8.5, 9, 9.5, 10$ and 17 for the cubic lattice. Further, values of $\varTheta$ for the hcp lattice were determined in the range $aR^* = 8, 8.5, \ldots, 16.5$. These values are listed below in Table II. The equilibrium value of R^* is again calculated from $\partial(U_{pot} + U_{zp})/\partial R^* = 0$; this gives, with (6) and (11)

$$0 = 6\frac{C_6}{R^{*7}} - 72\,(\varTheta - \varTheta')\,e^{a(1-R^*)} +$$

$$+ \frac{G(a)\,\varLambda^*\left[40\frac{C_8}{R^{*9}} + 24\left(\frac{1}{R^{*2}} + \frac{a}{R^*} - \frac{a^2}{2}\right)e^{a(1-R^*)}\right]}{\left[\left(\frac{a}{2} - \frac{1}{R^*}\right)24\,e^{a(1-R^*)} - 5\frac{C_8}{R^{*8}}\right]^{\frac12}}, \quad (12)$$

where $\varTheta' = \partial\varTheta/\partial(aR^*)$. Corner has given values of $\varTheta - \varTheta'$ for the fcc lattice; his results were extended to include the hexagonal lattice. Values of $\varTheta - \varTheta'$ are also given in Table II.

44

TABLE II

	Crystal functions for the fcc and hcp lattices			
aR*	fcc		hcp	
	Θ	$\Theta - \Theta'$	Θ	$\Theta - \Theta'$
8	1.024365	1.036574	1.024450	1.036677
8.5	19022	28334	19096	28428
9	14930	22086	14995	22173
9.5	11773	17314	11829	17389
10	9323	13640	9370	13703
10.5	7409	10790	7448	10843
11	5906	8567	5938	8612
11.5	4721	6824	4747	6861
12	3783	5451	3804	5481
12.5	3037	4365	3054	4389
13	2443	3503	2456	3522
13.5	1968	2818	1978	2832
14	1587	2267	1595	2280
14.5	1281	1828	1288	1838
15	1036	1476	1041	1483
15.5	838	1103	842	1198
16	678	965	681	969
16.5	—	—	552	—
17	445	632	—	635

§ 4. *Results for the Lennard Jones* $(s, 6)$ *potential.* The analysis was carried out for a Lennard Jones potential with the parameter s varying between 7 and 16, and for values of the zero-point energy parameter $\Lambda^* = 0.1, 0.2, \ldots 0.7$. The results for $\Lambda^* = 0$ can be found in the paper by K i h a r a and K o b a [4]). From eq. (10) the equilibrium value of r_0/R_0 is calculated for each set of parameter values s, Λ^*. The potential energy is obtained by

TABLE III

	Potential, zero-point and total energies of a fcc and a hcp lattice, for a Lennard Jones potential $(s, 6)$						
	$\Lambda^* = 0.3$						
s	U_{pot}/ε	U_{zp}/ε	U_{fcc}/ε	U_{pot}/ε	U_{zp}/ε	U_{hcp}/ε	relative difference $(\times 10^5)$
16	— 7.899357	1.150837	— 6.748520	— 7.900080	1.150818	— 6.749262	—11.0
14	— 8.063533	1.067337	— 6.996196	— 8.064274	1.067319	— 6.996954	—10.9
12	— 8.493963	1.112427	— 7.381536	— 8.494757	1.112415	— 7.382342	—10.9
10	— 9.100284	1.096064	— 8.004220	— 9.101057	1.096060	— 8.004997	— 9.7
9	— 9.591565	1.094049	— 8.497516	— 9.592297	1.094052	— 8.498245	— 8.6
8	—10.32791	1.101872	— 9.226041	—10.32856	1.101882	— 9.226681	— 6.9
7	—11.52693	1.129351	—10.39758	—11.52741	1.129369	—10.39804	— 4.4
	$\Lambda^* = 0.7.$						
16	— 7.152369	1.643276	— 5.509093	— 7.153209	1.643473	— 5.509736	—11.7
14	— 7.087318	1.394353	— 5.692965	— 7.088160	1.394506	— 5.693654	—12.1
12	— 7.991562	1.891553	— 6.100009	— 7.992460	1.891722	— 6.100738	—12.0
10	— 8.698353	2.013700	— 6.684653	— 8.699060	2.013681	— 6.685379	—10.9
9	— 9.234731	2.052129	— 7.182602	— 9.235421	2.052125	— 7.183296	— 9.7
8	—10.01250	2.128352	— 7.834150	—10.01314	2.128365	— 7.884775	— 7.9
7	—11.24837	2.244176	— 9.004198	—11.24888	2.244209	— 9.004669	— 5.8

substituting r_0/R_0 into (9) and (3) gives the zero-point energy. It appears that the values for the relative differences $(U_{hcp} - U_{fcc})/10^5|U_{fcc}|$ are negative and of the same order of magnitude for all values of Λ^* considered. (between 0 and 0.7). As an illustration, table III gives the numerical results for $\Lambda^* = 0.3$ and for $\Lambda^* = 0.7$.

A negative relative difference means that the hcp lattice is more stable than the fcc crystal structure. Note that the differences are only of the order of one tenthousandth of the cohesive energy.

The conclusion which can be drawn from the analysis is that for an additive Lennard Jones potential $(s, 6)$ and any plausible value of the parameter s, the hexagonal lattice is more stable than the cubic structure, including zero-point energy effects. It is seen also that the nonadditive third-order effect in the London forces [7]) is of the same order of magnitude as the differences in the last column of the tables above, but this time in favor of the face centered cubic lattice.

§ 5. *Results for the modified Buckingham potential* $(a, 6)$. The numerical solution of (12) is much more laborious than for the corresponding expression (10) in case of a Lennard Jones potential. We therefore follow another procedure: eq. (12) and the corresponding equation for the hcp lattice are solved numerically for Λ^*, with a and R^* as parameters. For each value of a, R^* is estimated such that the solution for Λ^* lies within the range 0, ..., 0.7. This process is repeated till the density of values for Λ^* in the range is sufficient. The crystal functions Θ and $\Theta - \Theta'$ must be interpolated from the values given in the table (II). The values for R^* and Λ^* are then substituted into (11) and (6) to give the potential and zero-point energies of the fcc and the hcp lattices. The differences in R^* between the two structures, belonging to the same a and Λ^*, are not larger than one unit in the fifth decimal place; these differences have been neglected in the results below. The same qualitative results are obtained as with the Lennard Jones potential field; the relative differences in the total energies are negative and of the same order

TABLE IV

		Potential, zero-point and total energies of a f cc and a hcp lattice, for a modified Buckingham potential $(a, 6)$						
a	Λ^*	U_{pot}/ε	U_{zp}/ε	U_{fcc}/ε	U_{pot}/ε	U_{zp}/ε	U_{hcp}/ε	relative difference ($\times 10^5$)
16	0.299023	—7.960695	1.133036	—6.827660	—7.961577	1.133129	—6.828448	—11.5
	0.647619	—7.455090	1.765990	—5.689101	—7.455755	1.766198	—5.689557	— 8.0
14	0.258588	—8.248339	0.974111	—7.274228	—8.249283	0.974174	—7.275108	—12.1
	0.651189	—7.824520	1.849335	—5.975186	—7.825406	1.849469	—5.975937	—12.6
12	0.286972	—8.653540	1.017860	—7.635680	—8.654535	1.017904	—7.636632	—12.5
	0.719104	—8.272319	2.007954	—6.264365	—8.273294	2.008035	—6.265260	—14.3
10	0.308336	—9.539230	1.021824	—8.517406	—9.540053	1.021766	—8.518287	—10.3
	0.742324	—9.290628	2.110625	—7.180003	—9.291580	2.110627	—7.180953	—13.2

of magnitude (one tenthousandth of the cohesive energy) for all values of Λ^* in the range between 0 and 0.7. As was stated before we have not included values for a lower than 10 in the analysis, because for $a < 10$ the modified Buckingham potential becomes physically unrealistic. In addition, the experimental values of a for the rare gases lie between 12 and 15, so that we are only interested in stability conditions for $a > 10$. In table IV the results are illustrated for Λ^* in the neighborhood of 0.3 and 0.7.

The results show that for an additive "modified" Buckingham potential the hcp lattice is again somewhat more stable than a face centered cubic structure. The difference is of the same order of magnitude with or without zero-point energy, and also of the same order as for an additive Lennard Jones $(s, 6)$ potential form.

§ 6. *Discussion of the results.* The results of this analysis confirm the conclusions reached by K i h a r a and K o b a [4]), who neglected the influence of zero-point energy on the stability of the crystals, that both for an additive Lennard Jones $(s, 6)$ potential and a modified Buckingham potential $(a, 6)$, with $a \geqslant 10$, the hcp lattice if slightly more stable than the fcc structure for the heavy rare gases neon, argon, krypton and xenon. Although the difference in energy between the two structures is only of the order of one tenthousandth of the cohesive energy, the constancy of this difference over the wide range of values for the parameters s and a seems to indicate that an explanation for the crystal structures of the heavy rare gases *cannot* be obtained by a slight adjustment in the interaction potential between two atoms or molecules, but will probably lie in the consideration of manybody forces, i.e. in a deviation from the assumption of additivity. As was mentioned before (§ 1) the preferrence for the fcc structure may then arise from differences in *symmetry* between the crystals (e.g. a compression of the electron clouds is more favorable in the fcc lattice than in the hcp crystal, see P r i n s [6])) or primarily from a difference in *density* (nonadditive contributions weaken the attractive field and therefore tend to favor the cubic structure) *).

In connection with the difference in cohesive energy between a fcc and a hcp structure it is of interest to mention that the diatomic molecules nitrogen

---

*) Further evidence that intermolecular forces are not quite additive may be obtained from measurements on the coefficients of elasticity of molecular crystals (see e.g. HCB, Chapter 13, p. 1035, footnote). Nonadditive contributions will, in general, cause a deviation from the Cauchy-relations (for an extensive discussion of the Cauchy-relations, see especially I. Stakgold, Q. Appl. Math. **8**, (1950) 169). However, as was pointed out by L. S a l t e r (Phil. Mag. **45**, (1954) 360), deviations from the Cauchy-relations do not necessarily invalidate the hypothesis of two-body forces (for solids of the inert gases), since these deviations may also be due to the influence of zero-point energy.

It would be important if the elasticity of molecular crystals could be measured accurately at very low temperatures; the experimental difficulties, are, however, very great. (B a r k e r et al., Phil. Mag. **77** (1953) 1182.

and carbon monoxide crystallize in the face centered cubic lattice at very low temperatures. At higher temperatures (35.61°K for nitrogen and 61.57°K for carbon monoxide) *a transition occurs to the hexagonal lattice*. In this case, however, is the stability of the fcc structure at the lowest temperatures determined primarily by a preferred orientation of the molecular axes. This problem will be discussed in more detail in chapter VI.

REFERENCES

1) C o r n e r, J., Trans. Far. Soc. **44** (1948) 914.

2) M a s o n, E. A. and R i c e, W. E., J. Chem. Phys. **22** (1954) 843.

3) L e n n a r d J o n e s, J. E. and I n g h a m, A. E., Proc. Roy. Soc. (London) A **107** (1925) 636.

4) K i h a r a, T. and K o b a, S., J. Phys. Soc. Japan **7** (1952) 348.

5) K i h a r a, T., Revs. Modern Phys. **25** (1953) 831.

6) P r i n s, J. A., D u m o r é, J. M. and L i e T i a m T j o a n, Physica **18** (1952) 307.

7) A x i l r o d, B. M., J. Chem. Phys. **19** (1951) 724.

8) C o r n e r, J., Trans. Far. Soc. **35** (1939) 711.

9) G o e p p e r t-M a y e r, M. and K a n e, G., J. Chem. Phys. **8** (1940) 642.

10) R i c e, W. E. and H i r s c h f e l d e r, J. O., J. Chem. Phys. **22** (1954) .

CONTRIBUTION TO THE ANALYSIS OF MOLECULAR INTERACTIONS IN COMPRESSED NITROGEN AND CARBON MONOXIDE

PART I. THE MOLECULAR FIELD IN THE HIGH DENSITY GAS STATE

§ 1. *Introduction.* Although nitrogen and carbon monoxide differ very markedly in chemical behaviour, the physical properties of these gases are almost the same. This is because they have the same total number of electrons, almost the same molecular weight and the equilibrium distance between the nuclei differs only very little. Information from molecular spectra shows that the molecules have identical ground states; further the rotational and vibrational constants are remarkably alike. The centers of symmetry of the positive and negative charges in carbon monoxide do not coincide but the resulting dipole is very small (0.11 D). In the region of high densities differences persist (boiling point, melting point, density of the liquid and solid states, etc.). So far, no theoretical treatment of the differences has been attempted based on a detailed study of the intermolecular field.

Accurate experimental values of all thermodynamic functions of compressed nitrogen and carbon monoxide have become available as a result of highly refined pVT-measurements. These values will be the experimental basis for the comparison of the two gases.

An assumption must be made about the form of the intermolecular field between two molecules. If the charge distributions of two interacting molecules do not overlap appreciably, then the Coulomb interaction can be developed into multipole fields. Applying perturbation theory, the first order interaction energy is the Coulomb interaction between the various permanent multipoles; the second order energy consists of interactions between induced multipoles. This expression in its general form is too complicated, therefore some restriction must be made at the outset. It is well known that for spherical molecules the interaction potential is in many cases adequately represented by the Lennard-Jones formula:

$$\Phi(r) = 4\varepsilon[(\sigma/r)^{12} - (\sigma/r)^6], \tag{1}$$

where ε and σ are parameters and r is the distance between the centers of the two molecules. For non-spherical molecules we have to supplement this expression by a term due to the interaction between the permanent multipoles and by a correction term due to the anisotropy in the induced multipole field. These correction terms will be denoted by Φ_A. Writing the Lennard-Jones potential as Φ_{L-J} we have, formally:

$$\Phi = \Phi_{L-J} + \Phi_A. \tag{2}$$

This approach has the advantage that it is possible to determine the parameters ε and σ from experimental second virial coefficients. Physically speaking this means of course that in the low density region where the second virial coefficient is predominant, the average distance between the molecules is so large that the orientation part of the potential field can be neglected.

The thermodynamic properties of the two gases have now to be discussed on the basis of equation (2). In section 2, experimental values of the internal energy, internal entropy and specific heat at constant volume, in the density region 0–600 Amagat and for temperatures between 0°C and 150°C are compared. Although the thermodynamic functions can in general not be written as the sum of a "$L-J$"part and a "A" part, the assumption of additivity holds at high densities in good approximation (section 3a). This assumption makes it possible to determine the differences in thermodynamic properties between the two gases due to different values of the parameters ε and σ, by an application of the theorem of corresponding states (section 3b). The calculations in sections 4, 5 are restricted to the potential field Φ_A (orientation effects at high densities). In sections 4a, b the effect of dipole orientation and dipole induction for carbon monoxide at high densities is calculated. The effect of the anisotropy in the induced dipole field will be treated in section 5. The calculations are all based on the assumption that the intermolecular field may be written as a sum of interactions between isolated pairs of molecules (additivity of intermolecular forces). Deviations from this assumption at high densities will be discussed in detail in section 6.

§ 2. *Thermodynamic properties and intermolecular parameters of nitrogen and carbon monoxide.* In figs. 1–3 experimental data on the internal energy, internal entropy, and specific heat at constant volume are plotted as functions of temperature, for a series of densities. In figs. 4 and 5 the difference between the internal energy respectively entropy of CO and N_2 is given as a function of temperature for different densities. M i c h e l s, W o u t e r s and D e B o e r [1] [2] determined the pVT-data and calculated the thermodynamic properties of nitrogen; for carbon monoxide the pVT-values were determined by M i c h e l s, L u p t o n, W a s s e n a a r and D e G r a a f f [3] and the thermodynamic values calculated by M i c h e l s, L u n b e c k and W o l k e r s [4]. The experimental data show that the differences between N_2 and CO increase with increasing density and decrease with rising temperature, at least at high temperatures.

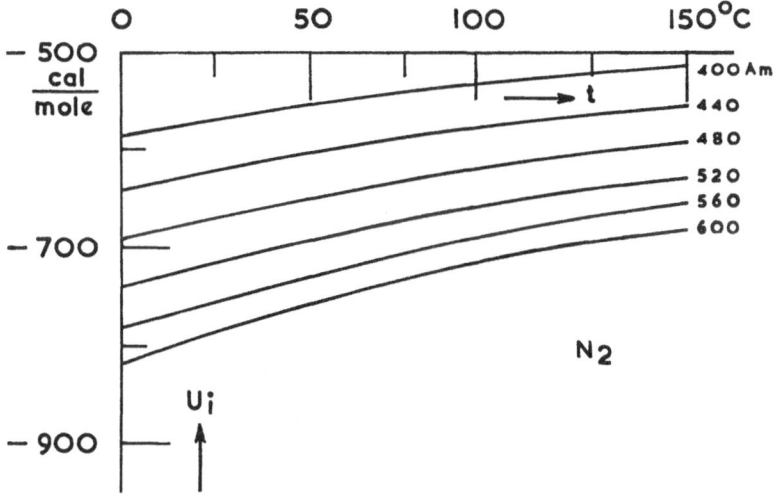

Fig. 1a. Internal energy of N_2 as a function of the temperature for different densities

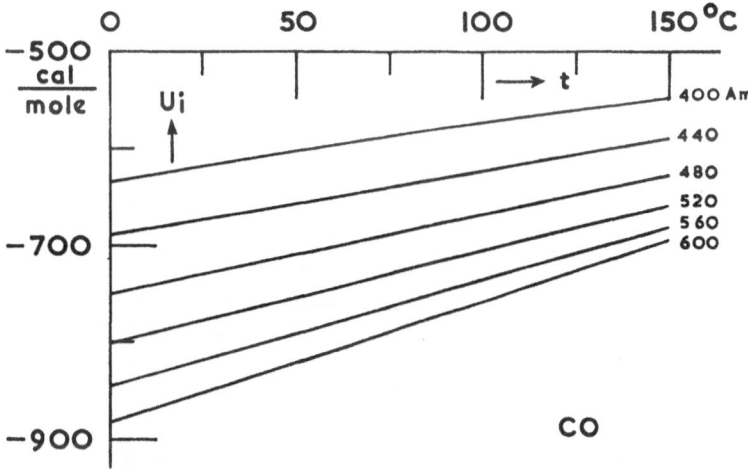

Fig. 1b. Internal energy of CO as a function of the temperature for different densities.

From values of the second virial coefficients the intermolecular parameters ε and σ can be evaluated. The most accurate values seem to be:

	CO	N_2
$\varepsilon \times 10^{16}$ ergs	138.2	131.3
σ Å	3.77	3.71
$\varepsilon_{CO}/\varepsilon_{N_2}$	1.052	
$(\sigma_{CO}/\sigma_{N_2})^3$	1.051	

If orientation is important at high densities, then it must be reflected in the internal energy of the system. Further, the system will have a lower entropy, since the molecules are restricted in their freedom of rotation. The specific heat at constant volume should be higher than in the case of free rotation, due to the additional energy required to free the molecules from their

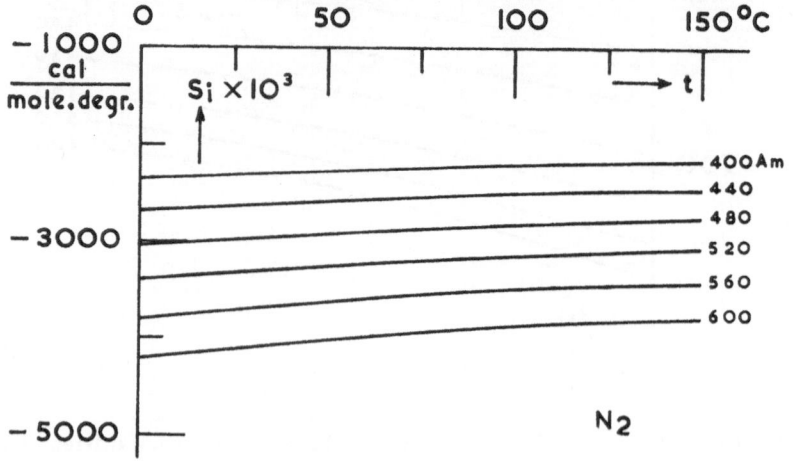

Fig. 2a. Internal entropy of N_2 as a function of the temperature for different densities.

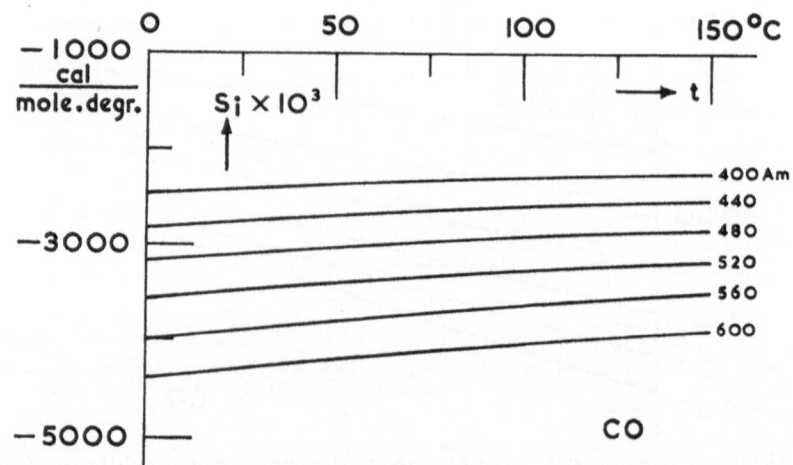

Fig. 2b. Internal entropy of CO as a function of the temperature for different densities

rotational restrictions. Since, however, the thermodynamic properties depend on the complete potential function (2), we must carry through a detailed analysis to discern the effect of multiple orientation in the dense state.

§ 3a. *Cell theory and the theorem of corresponding states.* The statistical problem of calculating thermodynamic functions of compressed gases from

52

the interaction potential between two molecules is in practice almost impossible to solve. Approximate methods have been developed for low densities, in the form of a virial expansion, and for very high densities, based on a cell model. For the high density region calculations were performed by H i r s c h f e l d e r c.s. [5]), using the free volume theory as developed by L e n-

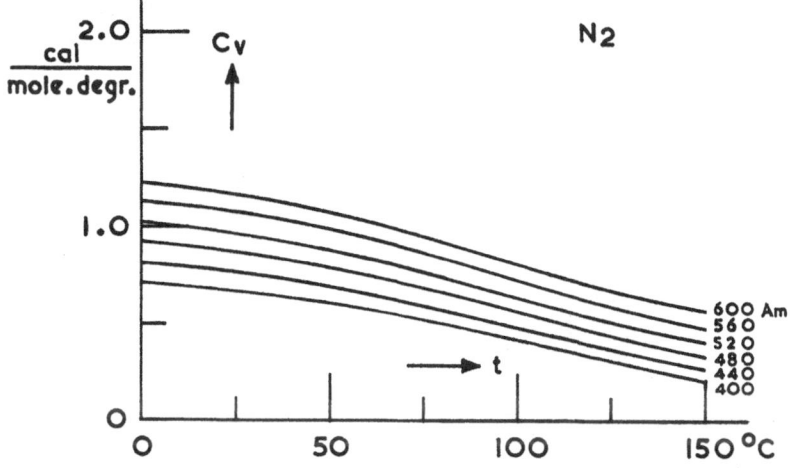

Fig. 3a. Specific heat at constant volume of N_2 as a function of the temperature for different densities.

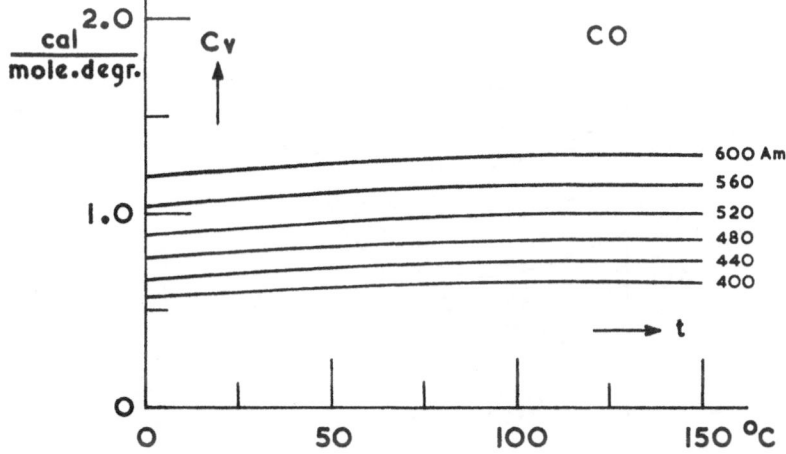

Fig. 3b. Specific heat at constant volume of CO as a function of the temperature for different densities.

n a r d - J o n e s and D e v o n s h i r e [6]). As can be expected on this basis, the agreement with experiment is best at the highest densities; yet some systematic deviations from the data are found. For argon, for instance, the calculated internal energy is lower (more negative) than the experimental values, the difference being of the order of ten percent at a density of 600 Amagat.

For the calculations on nitrogen the potential function was assumed to be of the Lennard-Jones form; the authors neglected therefore effects of multiple orientation. Then it is found that the calculated internal energy is higher (less negative) than the experimental values; the difference is of the order of ten percent at a density of 480 Amagat. The authors conclude that the deviation

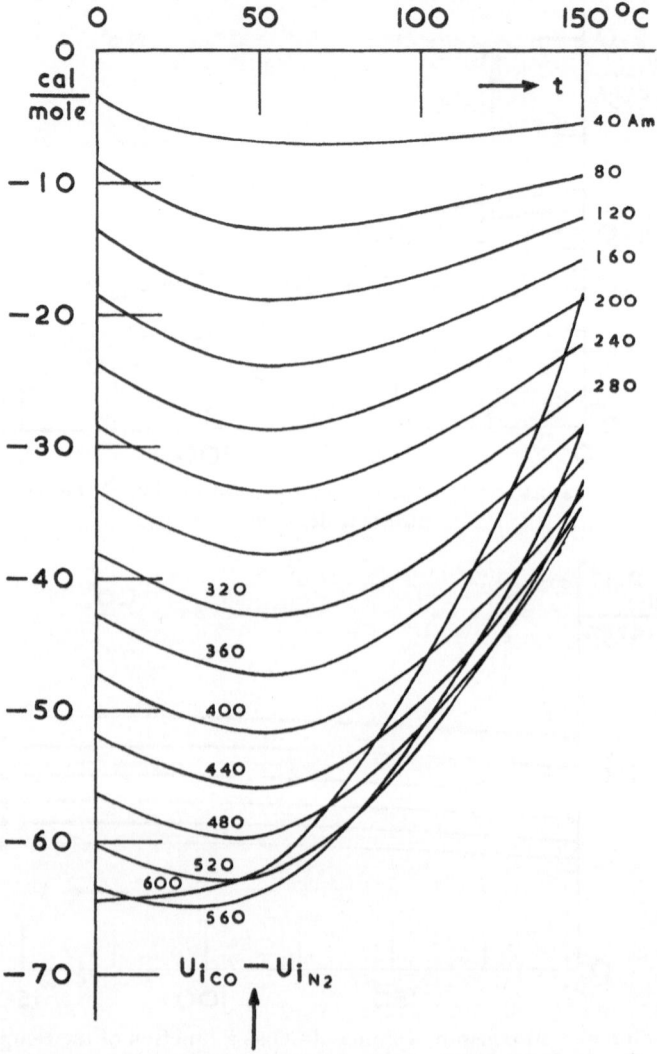

Fig. 4. The difference between the internal energy of CO and N_2 as a function of the temperature for different densities.

may be due to the non-spherical symmetry of the intermolecular field between two nitrogen molecules.

For the comparison between nitrogen and carbon monoxide, cell theoretical methods are too inaccurate to be used directly for the calculation of thermodynamic properties. Therefore we will follow another procedure. It is wel known that, if the potential field between two molecules is purely of the Lennard-Jones form, for instance, the thermodynamic functions in "reduced" form are unique functions of "reduced" variables (theorem of corresponding states). Then, if we know the ratio of ε and σ for two gases, we can calculate

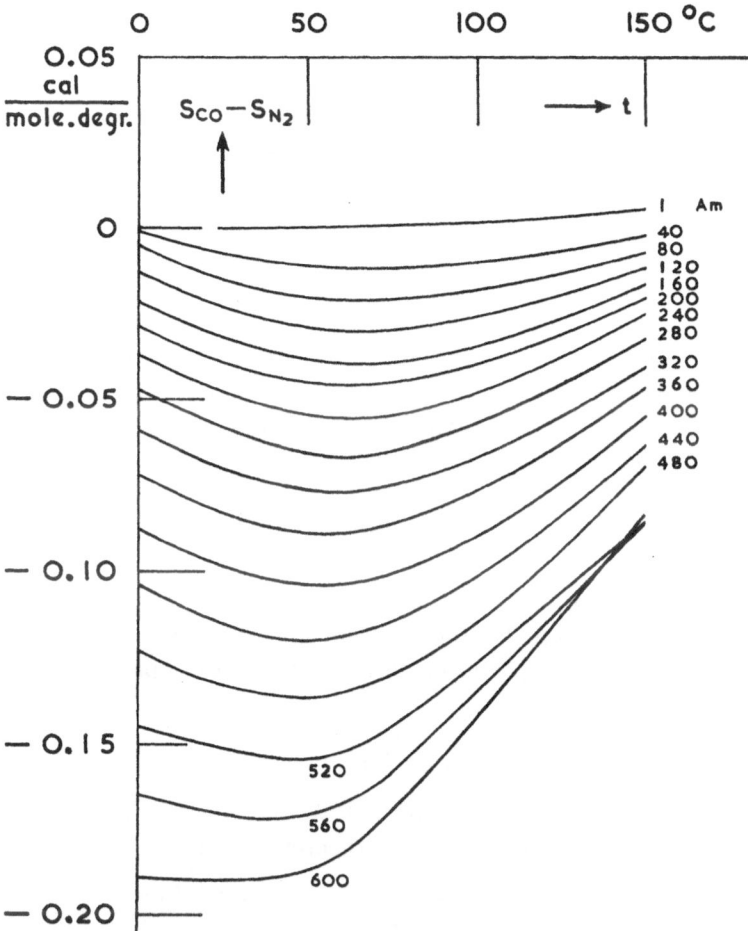

Fig. 5. The difference between the entropy of CO and N_2 as a function of the temperature for different densities.

directly the difference in the thermodynamic functions, starting from the experimental values of one of the two. If the potential field between two molecules is not spherical the theorem of corresponding states is not applicable in this form. However, if the thermodynamic functions were additive in the different parts of the potential field, then the theorem of corresponding states may be applied to the corresponding "Lennard-Jones" part of the thermodynamic functions. It is easily seen that this assumption of additivity in the thermodynamic functions is not valid in general. This is because the thermodynamic functions are ensemble averages statistically and the probability density for any micro-configuration depends on the complete potential field. Only if the potential field consists of two parts, one depending only on the relative positions of the centers of the molecules and the other part depending only on the relative orientations of the molecules, then the assumption of additivity holds.

Since cell theoretical calculations show that the Lennard-Jones part of the potential field is most important at all densities considered and since our main interest for the comparison between nitrogen and carbon monoxide lies in the region of high densities (between 400 and 600 Amagat), we make the following assumption: In the region of high densities the thermodynamic properties may be calculated as if the molecules were situated at the centers of their respective cells as far as the "orientational" part of the potential field (Φ_A) is concerned.

With this approximation the total partition function of the system is the product of two parts, translational and rotational, and the thermodynamic functions are additive in the two parts of the potential field *). For instance the Helmholtz free energy, F, can then be written as:

$$F = F_{L-J} + F_A \tag{3}$$

The same is true for all other thermodynamic functions.

§ 3b. *Application to nitrogen and carbon monoxide.* With the approximation outlined in section 3a, the internal energy of nitrogen and carbon monoxide is a sum of two parts, one term due to the $L-J$ part of the potential field and one part due to Φ_A:

$$U_{exp} = U_{L-J} + U_A$$

If we define a "reduced temperature" $T^* = kT/\varepsilon$ and a "reduced volume" $V^* = V/N\sigma^3$, then we obtain from the theorem of corresponding states:

$$(U^*_{(L-J)})_{CO} = (U^*_{L-J})_{N_2} \qquad \text{(at the same } T^* \text{ and } V^*) \tag{4}$$

with $U^* = U/N\varepsilon$

This gives:

$$(U_{L-J})_{CO} = (\varepsilon_{CO}/\varepsilon_{N_2}) \cdot (U_{L-J})_{N_2} \qquad \text{(at the same } T^* \text{ and } V^*)$$

and we obtain the expression:

$$U_{CO} - (\varepsilon_{CO}/\varepsilon_{N_2})\, U_{N_2} = U_{A\,CO} - (\varepsilon_{CO}/\varepsilon_{N_2})\, U_{A\,N_2} \text{ (at the same } T^* \text{ and } V^*) \tag{5}$$

---

*) J. A. P o p l e, ref. 9, used the same approximate method for the effect of dipole orientation and induction on the cohesive energy of polar liquids. An extensive discussion of this method and further applications can be found in a paper by the same author: J. A. P o p l e, Discussions of the Far. Soc. No. **15**, 35, (1953).

The quantities on the left hand side are known from experiments. In this way we obtain information on the differences between nitrogen and carbon monoxide due to multiple orientation. In fig. 6, the difference $U_{\Delta CO} - (\varepsilon_{CO}/\varepsilon_{N_2}) U_{\Delta N_2}$ is plotted as a function of temperature, for different densities *). Note that the difference is positive at all temperatures. This is the most accurate information we can obtain on the difference in thermodynamic properties. It has been tacitly assumed that the values of the parameters ε and σ are the same at high densities as at low densities. This assumption is related to the "additivity of intermolecular forces" and will be discussed in detail in section 6.

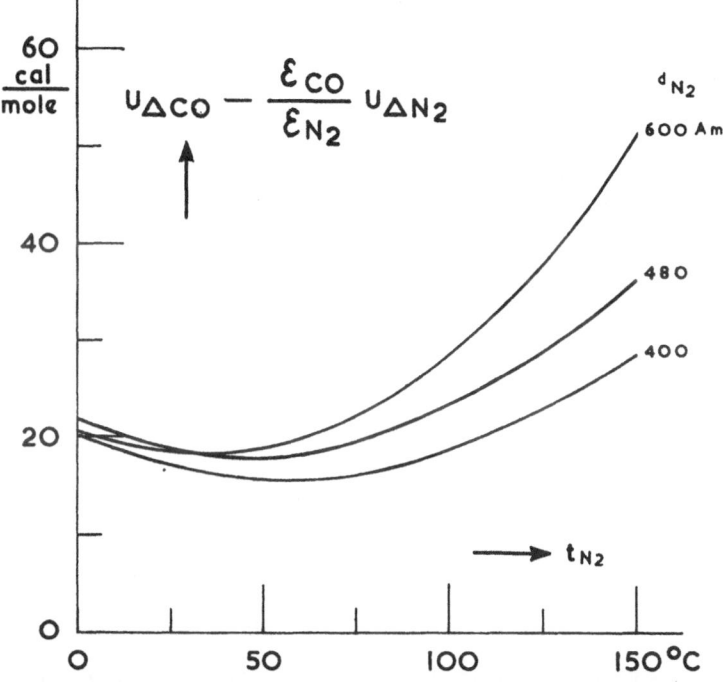

Fig. 6. $U_{\Delta CO} - (\varepsilon_{CO}/\varepsilon_{N_2}) U_{\Delta N_2}$ as a function of the temperature.

§ 4a. *The orientation effect of the dipoles of carbon monoxide at high densities.*
First we discuss the difference in internal energy between the two gases, due to the orientation of the permanent dipoles of carbon monoxide. With the approximation of section 3, we have to evaluate the partition function for an assembly of N point dipoles, with dipole vectors $\mu_i = \mu . e_i$; $i = 1, 2, \ldots N$. μ is the dipole moment and e is a unit vector. Each of the dipoles is considered fixed at the center of its cell. We must determine the integral:

$$Z_N^{(dip)} = \int \ldots \int \exp\left(-\Phi^{(dip)}/kT\right) d\, e_N, \tag{6}$$

*) For each t_{N_2} and d_{N_2} corresponding temperatures and densities t_{CO} and d_{CO} were determined. With the help of the experimental data for U_{N_2} at t_{N_2} and d_{N_2} and of U_{CO} at t_{CO} and d_{CO}, the graph in fig. 6 was evaluated.

where

$$\Phi^{(\text{dip})} = \tfrac{1}{2} \Sigma_{i,j} \, \Phi_{ij}^{(\text{dip})},$$

$$\Phi_{ij}^{(\text{dip})} = -\mu^2 \left[3(\mathbf{e}_i \cdot \mathbf{R}_{ij})(\mathbf{e}_j \cdot \mathbf{R}_{ij})/R_{ij}^5 - (\mathbf{e}_i \cdot \mathbf{e}_j)/R_{ij}^3 \right] \tag{7}$$

and R_{ij} is the distance between the centers of cells i and j. The integration is to be carried out subject to the restraint:

$$(\mathbf{e}_j \cdot \mathbf{e}_j) = 1 \text{ for } j = 1, 2, \ldots N. \tag{8}$$

This is the classical model; the evaluation of (6) has not been performed with the natural restriction on $\mathbf{e}$. Approximate methods, however, are avaiblable. One is due to B e r l i n , T h o m s e n [7]) and L a x [8]) and is valid for all temperatures and for all values of the dipole moment. In this method the restriction (8) is replaced by the much weaker condition:

$$\Sigma_{j=1}^{N} (\mathbf{e}_j \cdot \mathbf{e}_j) = N \tag{8'}$$

after which the main contribution to the integral (6) is determined by means of a saddle point method. Another method was developed by P o p l e [9]). Here the integrand is developed as a power series in $\Phi^{(\text{dip})}/kT$; the series converges rapidly if $\mu^2/vkT \ll 1$ (v is the available volume per molecule). In this region of small dipole moments and high temperatures the two methods give the same results. The dipole moment of CO is about $0.1 \text{ D} = 10^{-19}$ e.s.u. At a density of 600 Amagat the volume per molecule is of the order of $60 \cdot 10^{-24} \text{ cm}^3$. For $T = 300°K$ we obtain for μ^2/vkT a value of about $4 \cdot 10^{-3}$. We follow Pople's method and write:

$$Z_N^{(\text{dip})} = \Sigma_{p=0}^{\infty} (-1)^p/p! \int \ldots \int [\Phi^{(\text{dip})}/kT]^p \, d\omega_N =$$
$$= (4\pi)^N (1 + \Sigma_{p=1}^{\infty} M_p), \tag{9}$$

where

$$M_p = (4\pi)^{-N} \cdot (-1)^p/p! \int \ldots \int [\Phi^{(\text{dip})}/kT]^p \, d\omega_N$$

and $\int d\omega_N$ has been written for the integration over the angular coordinates of the molecules. In terms of the angular coordinaties θ, φ of two molecules, we have:

$$\Phi_{ij}^{(\text{dip})} = - (\mu^2/R_{ij}^3)(2 \cos \theta_i \cos \theta_j - \sin \theta_i \sin \theta_j \cos \varphi_{ij}).$$

The term M_1 gives the approximation in which each dipole is considered as independent of the others; therefore M_1 is zero. The higher moments M_2, $M_3 \ldots$ arise from multiple orientation effects between the dipoles.

For the Helmholtz free energy of the system of dipoles, $F_N^{(\text{dip})}$, one obtains an expression of the form:

$$F_N^{(\text{dip})} = - kT \ln Z_N^{(\text{dip})} = - NkT \, \Sigma_{r=2}^{\infty} c_r \cdot a^r, \tag{10}$$

where

$$a = \mu^2/vkT.$$

The coefficients c_r depend on the orientation function $f(\theta_N, \varphi_N)$ in the interaction potential between the molecules and on the type of lattice.

For a face centered cubic array and dipole orientation froces $c_2 = 1.2045$; $c_3 = -0.638$; $c_4 = -0.108$ (P o p l e, *loc. cit.*). Then the internal energy of the dipole lattice is:

$$U_N^{(dip)} = -\partial \ln Z_N^{(dip)}/\partial\, 1/kT = -2\, Nc_2\mu^4/v^2 kT \simeq -2 \times 10^{-2} \text{ cal/mole}$$

at $T = 300°K$ and a density of 600 Amagat. For the specific heat of rotation we have:

$$c_v^{(dip)} = 0.02/300 \simeq 10^{-4} \text{ cal/mol. degree.}$$

These values are much too small to account for the differences between the two gases.

§ 4b. *The dipole induction effect at high densities for carbon monoxide.* For the evaluation of the induction effect due to the permanent dipoles of carbon monoxide we consider again a lattice, consisting of point dipoles. Each dipole is fixed at the center of its cell (fixed with respect to its position in the cell, not with respect to its orientation). The partition function for the induction effect is:

$$Z_N^{(ind)} = \int \ldots \int \exp\left(-\Phi^{(ind)}/kT\right) d\omega_N. \tag{11}$$

(Strictly speaking we are not allowed to evaluate the dipole orientation and induction effects separately. If, however, the two effects differ in value by an order of magnitude, the present method is a good approximation). We have:

$$\Phi^{(ind)} = \Sigma_{i=1}^N \Phi^{(i)}, \tag{12}$$

where $\Phi^{(i)}$ is the induced potential at the center of cell i due to all the other dipoles in the lattice.

$$\Phi^{(i)} = -\tfrac{1}{2}a\, [\Sigma_{j=1}^{\prime N} \mathbf{F}_j^{(i)}]^2 \tag{13}$$

The prime in the summation means that $j \neq i$. $\mathbf{F}_j^{(i)}$ is the field strength at the center of cell i induced by dipole j, and a is the polarizability of the molecules. We have:

$$\mathbf{F}_j^{(i)} = -\mu e_j/R_{ij}^3 + 3\mu \mathbf{R}_{ij}(e_j.\mathbf{R}_{ij})/R_{ij}^5. \tag{14}$$

It has to be noted that the polarizability of diatomic molecules is anisotropic; however, we do not have to take this into account to estimate the order of magnitude of the induction effect. Further, we will not consider the change in the polarizability with the density of the system (see section 6). For the evaluation of the integral we follow the same method as used for the orientation effect, i.e. we expand the integrand as a power series in $\Phi^{(ind)}/kT$. The first term in the series for $\ln Z_N^{(ind)}$ is not zero in this case:

$$M_1 = (Na/2kT)\,(4\pi)^{-N} \int \ldots \int (\Sigma_{j=2}^N \mathbf{F}_j^{(1)})^2\, d\omega_N =$$
$$= (Na/2kT)\,(4\pi)^{-2}\,\mu^2\, \Sigma_{j=2}^N \int\int (1 + 3\cos^2\theta_{1j})/R_{1j}^6\, d\omega_1\, d\omega_j. \tag{15}$$

59

Here $\cos \theta_{1j} = (\mathbf{e}_j . \mathbf{R}_{1j})/R_{1j}$, i.e. θ_{1j} is the angle between the dipole vector of dipole j and $\mathbf{R}_{1j}$. Cross terms are zero on integration. Eq. (15) may be written as:

$$M_1 = N(a/v)\,(\mu^2/vkT)\,(\Sigma^N_{j=2}\,v^2/R^6_{1j}) \tag{16}$$

For carbon monoxide $a = 1.844 \times 10^{-24}$ cm^3 (ref. 10). Lattice sums of the type $\Sigma^N_{j=2}\,(R_0/R_{1j})^6$, where R_0 is the distance between nearest neighbors, have been evaluated by Lennard-Jones and Ingham[11]). For a face centered cubic lattice we have $R^3_0 = \sqrt{2}.v$; then we find $\Sigma^N_{j=2}(v^2/R^6_{1j}) = 7.2$. The internal energy due to the induction effect is:

$$U_N^{(ind)} = -\,\partial \ln Z_N^{(ind)}/\partial\,1/kT \simeq -\,0.8 \text{ cal/mole}$$

Higher terms need not be evaluated. Although the induction effect is much larger than the orientation effect, it is still much too small to account for the difference between nitrogen and carbon monoxide. The result of this section and of section 4a is that the differences in the thermodynamic properties between the two gases cannot be ascribed to an effect of the permanent dipoles of carbon monoxide.

§ 5. *Anisotropy of the dispersion forces.* If the charge distribution in a molecule is not spherical, then the dispersion forces are anisotropic. This effect was calculated by J. H. de Boer and Heller[12]), who considered the general expression for the dispersion forces for molecules having three perpendicular axes of symmetry. Massey and Buckingham[13]) evaluated the attractive field between two hydrogen molecules using the quantum mechanical approximation method as developed by Lennard-Jones. J. de Boer[14]) showed that the agreement of these calculations with experiment can be improved by a combination of the two methods.

We consider molecules having three perpendicular axes of symmetry; the axes are x (axis of rotational symmetry), y and z. The polarizabilities along these axes are a_x, a_y and a_z, respectively; for diatomic molecules $a_y = a_z$. If we introduce the following expressions for the average polarizability $\bar{a}$ and the anisotropy factor γ:

$$\bar{a} = (a_x + 2a_y)/3; \qquad \gamma = (a_x - a_y)/3\bar{a},$$

then the expression for the induced dipole interaction between two molecules, in terms of the parameters ε and σ, the distance between their centers r_{12}, the orientations θ, φ, and the anisotropy factor γ can be written as follows: (J. de Boer, *loc. cit.*)

$$E = -\,4\varepsilon\,(\sigma/r_{12})^6\,[1 - (1 - \tfrac{3}{2}\cos^2\theta_1 - \tfrac{3}{2}\cos^2\theta_2)\,\gamma -$$
$$-\{\tfrac{3}{2}\cos^2\theta_1 + \tfrac{3}{2}\cos^2\theta_2 - \tfrac{3}{2}(2\cos\theta_1\cos\theta_2 - \sin\theta_1\sin\theta_2\cos\varphi_{12})^2\}\,\gamma^2], \tag{17}$$

i.e. we have a correction due to the anisotropy:

$$\Phi_{12}^{(a)} = 4\varepsilon \, (\sigma/r_{12})^6 \, [(1 - \tfrac{3}{2}\cos^2 \theta_1 - \tfrac{3}{2}\cos^2 \theta_2) \, \gamma +$$

$$+ \{\tfrac{3}{2}\cos^2 \theta_1 + \tfrac{3}{2}\cos^2 \theta_2 - \tfrac{3}{2} (2 \cos \theta_1 \cos \theta_2 - \sin \theta_1 \sin \theta_2 \cos \varphi_{12})^2\} \, \gamma^2]. \quad (18)$$

(eq. (17) was derived on the assumption that the fundamental vibration frequencies parallel and perpendicular to the rotational axis of symmetry are equal). Using the approximation of section 3 we have to evaluate the partition function:

$$Z_N^{(a)} = \int \ldots \int \exp \left(- \Phi^{(a)}/kT\right) d\omega_N,$$

where

$$\Phi^{(a)} = \tfrac{1}{2} \Sigma_{i,j}^N \, \Phi_{ij}^{(a)} = 2\varepsilon\sigma^6 \Sigma_{i,j}^N \, \beta_{ij}/R_{ij}^6 \quad (19)$$

and

$$\beta_{ij} = (1 - \tfrac{3}{2}\cos^2 \theta_1 - \tfrac{3}{2}\cos^2 \theta_2) \, \gamma + \{\tfrac{3}{2}\cos^2 \theta_1 + \tfrac{3}{2}\cos^2 \theta_2 -$$

$$- \tfrac{3}{2} (2 \cos \theta_1 \cos \theta_2 - \sin \theta_1 \sin \theta_2 \cos \varphi_{12})^2\} \, \gamma^2.$$

We follow the same procedure as in sections 4a, b. The first term in the series for $\ln Z_N$ is again M_1:

$$M_1 = - N \, \Sigma_j' \, (\sigma^6/R_{1j}^6) \, 2 \, (\varepsilon/kT) \, (4\pi)^{-2} \int\!\!\int \beta_{1j} \, d\omega_1 \, d\omega_j$$

where the prime indicates that $j \neq 1$ in the summation. When the integration is carried out it is found that M_1 is zero, just as for the orientation effect of the permanent dipoles of carbon monoxide. The next term in the series is M_2:

$$M_2 = (4\pi)^{-N} \int \ldots \int (\Sigma_{i>j} \, \Phi_{ij}^{(a)}) \, (\Sigma_{k>l} \, \Phi_{kl}^{(a)}) \, d\omega_N/2(kT)^2$$

The only non-vanishing combinations $(i, j; k, l)$ are of the form $(i, j; j, k)$ where $j \neq i$, $k \neq j$. They reflect interactions between three $(k \neq i)$ or two $(k = i)$ molecules. One obtains:

$$M_2 = (4\pi)^{-N} . N \int \ldots \int \{\Sigma_{i \neq k} \, \Phi_{1j}^{(a)} \, \Phi_{1k}^{(a)} + \tfrac{1}{2} \Sigma_j \, \Phi_{1j}^{(a)2}\} \, d\omega_N/2(kT)^2 \quad (20)$$

For $\ln Z_N^{(a)}$ we have in this approximation:

$$\ln Z_N^{(a)} = (4\pi)^{-N} . N \, (\varepsilon/kT)^2 \, (\sigma/R_0)^{12} \int \ldots \int [8 \, \Sigma_{j \neq k} \, (R_0/R_{1j})^6 \, (R_0/R_{1k})^6 \, \beta_{1j} \beta_{1k} +$$

$$+ 4 \, \Sigma_j \, (R_0/R_{1j})^{12} \, \beta_{1j}^2] \, d\omega_N. \quad (21)$$

Integration over the orientations has the result:

$$(4\pi)^{-N} \int \ldots \int \beta_{1j} \beta_{1k} \, d\omega_N = \gamma^2/5$$

$$(4\pi)^{-N} \int \ldots \int \beta_{1j}^{2j} \, d\omega_N = 2\gamma^2/5 + 19\gamma^4/25. \quad (22)$$

The values of the lattice sums $\Sigma_j' \, (R_0/R_{1j})^6$ and $\Sigma_j' \, (R_0/R_{1j})^{12}$ can be taken from the paper by L e n n a r d-J o n e s and I n g h a m (*loc. cit.*); for a face centered cubic array they are 14.4 and 12.1, respectively. To perform the first sum in (21), we write:

$$\Sigma_{j \neq k}' \, (R_0/R_{1j})^6 \, (R_0/R_{1k})^6 = (\Sigma_j' \, R_0^6/R_{1j}^6)^2 - \Sigma_j^1 \, R_0^{12}/R_{1j}^{12} = 195.3$$

with the values given before. To estimate the order of the anisotropy effect, we use for the two gases $\varepsilon/k = 100$ degr.; $\sigma = 3.7$ Å; $\gamma = 0.17$. At a density of 600 Amagat and a temperature of 300 °K, $\varepsilon/kT = {}^1/_3$; $(\sigma/R_0)^{12} = 0.16$. The internal energy is in this approximation:

$$U_N^{(a)} = - \partial \ln Z_N^{(a)}/\partial \, 1/kT = - 2kT \ln Z_N^{(a)} = - 170 \text{ cal/mole} \qquad (24)$$

At a density of 480 Amagat and a temperature of 0°C this value is about -70 cal/mole. The third term in the series for $\ln Z_N^{(a)}$ is M_3:

$$- M_3 = (4\pi)^{-N} \int \dots \int (\Sigma_{i>j} \, \varPhi_{ij}^{(a)}) \, (\Sigma_{k>l} \, \varPhi_{kl}^{(a)}) \, (\Sigma_{m>n} \, \varPhi_{mn}^{(a)}) \, \mathrm{d}\omega_N/6(kT)^3$$

The terms which do not vanish on integration are of the form $(ij, jk; kl)$, with $i \neq j$, $j \neq k$, $k \neq l$. Combinations with five or six different molecules are zero.

The sum $\Sigma_{i,j,k,l} \, \varPhi_{ij}^{(a)} \, \varPhi_{jk}^{(a)} \, \varPhi_{kl}^{(a)}$, ($i$, j, k, l are four different molecules), gives the largest contribution to the integral; terms with γ^3 and γ^5 are zero and the first non-vanishing contribution is proportional to γ^4. Evaluation of this term shows that the corresponding contribution to the internal energy is less than ten procent of the second term and has the same sign. For the comparison between nitrogen and carbon monoxide, we determine the ratio of the anisotropy effect with the help of eqs. (22) and (24):

$$U_{CO}^{(a)}/U_{N_2}^{(a)} = (\varepsilon_{CO}/\varepsilon_{N_2})^2 \, (\sigma_{CO}/\sigma_{N_2})^{12} \, (\gamma_{CO}/\gamma_{N_2})^2$$

at the same temperature and density. It is also possible to evaluate the difference $U_{CO}^{(a)} - (\varepsilon_{CO}/\varepsilon_{N_2}) \, U_{N_2}^{(a)}$ at the same reduced temperature and density; it appears that the values are positive but somewhat too small to account for the experimental differences between nitrogen and carbon monoxide *).

The anisotropy is, however, of importance for the explanation of the differences between cell theoretical calculations and the experimental values of the internal energy for non spherical molecules (see also section 3). Using only the Lennard-Jones part of the potential field H i r s c h f e l d e r c.s. (*loc. cit.*) obtained for a density of 480 Amagat and a temperature of 0°C a value of $- 640$ cal/mol for the internal energy of nitrogen, whereas the experimental value is -695 cal/mol. The inaccuracy of cell theoretical methods does not permit a quantitative comparison but it can be stated that multiple orientation due to the anisotropy in the dispersion forces results in a considerable decrease in the internal energy of non spherical molecules at high densities.

§ 6. *Influence of the change in polarizability with pressure on the thermodynamic properties of compressed gases.* In the foregoing sections it was

*) For the anisotropy factor we find from:
L a n d o l t-B o r n s t e i n, Physikalisch Chemische Tabellen: $\gamma_{CO} = 0.167$; $\gamma_{N_2} = 0.176$
K.G. D e n b i g h, Trans. Far. Soc. **36**, 936, (1940): $\gamma_{CO} = 0.168$; $\gamma_{N_2} = 0.189$

repeatedly assumed that the molecular parameters ε and σ do not change with density, i.e. they have at all densities the same values as in the low density region from which they were determined experimentally. The inherent assumption is obviously valid if the potential field at all densities may be written as a sum of terms referring to isolated pairs of molecules. This statement is known as the "additivity of intermolecular forces" and will now be discussed in more detail [15]).

It was mentioned in section 3 that H i r s c h f e l d e r c.s. calculated the thermodynamic properties of argon using the free volume theory as developed by L e n n a r d - J o n e s and D e v o n s h i r e. The calculated internal energy is found to be more negative than the experimental values, the difference being of the order of ten percent at a density of 600 Amagat.

There are several possibilities for these differences. First of all, the average number of molecules around a caged molecule may not be the same as in the crystalline state but lower. A straightforward interpretation of this decrease of the coordination number is provided by a generalized free volume theory given by R o w l i n s o n and C u r t i s s [16]) [17]) [18]). Another possibility may be that deviations from additivity in the intermolecular field must be taken into account at such high densities.

The assumption of additivity is not valid for molecules which tend to associate (for example molecules with hydroxyl or amino groups) or for molecules which form hydrogen bonds. We restrict ourselves to forces between spherically symmetric molecules, especially argon, and we exclude from consideration the types of forces mentioned above.

First we consider the dispersion forces between spherically symmetric atoms or molecules. L o n d o n [19]) and M a r g e n a u [20]) have shown that van der Waals dispersion forces are additive in second order perturbation theory. A x i l r o d and T e l l e r [21]) [22]) applied third order perturbation theory to the dipole dispersion forces between neutral atoms. It was observed that this order reflects the interaction between triplets of atoms, giving rise to a non additive contribution. The nonadditivity decreases the attraction in the case of an equilateral configuration of the three atoms, and increases the attraction for a collinear array of the triplet. A x i l r o d summed the third order interaction for crystals of the heavy rare gases argon, krypton and xenon. The third order energy is positive, thus decreasing the attractive field between the molecules, and amounts to two to nine percent of the cohesive energy.

If the distances between the three atoms are so small that the wave functions overlap appreciably, then the zero order wave function must be made antisymmetric with respect to the exchange of electrons between different atoms. First order perturbation theory then gives the repulsive

---

*) P_λ permutes electrons between atoms **a** and **b**; λ is even or odd for even or odd permutations, respectively.

interaction between the atoms. From the calculations by P h. R o s e n [23]) it appears that the first order forces between three helium atoms are not equal to the sum of interactions between isolated pairs; this means that first order forces do not have the property of additivity. In the equilateral triangular configuration of the triplet the first order interaction is lower than calculated on the assumption of additivity, whereas in the collinear array the repulsion is higher than in the case of three isolated pairs. The error involved in neglecting nonadditivity is one percent or less for the equilateral triangle if the distance between the atoms is larger than 4.8 Bohr radii; for the colli-near array it is always smaller.

The nonadditivity calculations of R o s e n and A x i l r o d are straight-forward extensions of the evaluation of first and second order interactions between two atoms or molecules. It must be expected that for the calculation of second order interactions at high densities the simple-product type of wave function is a poor approximation and that therefore exchange terms should be taken into account. If of the three atoms a, b, c two are close together (a, b) and one is far apart (c), then the zero order wave function, using the valence bond method, can be written as *):

$$\Psi_0 = (\Sigma_\lambda (-1)^\lambda P_\lambda \Psi_a \Psi_b) . \Psi_c \qquad (25)$$

i.e. the dispersion forces between ab and c, as calculated on the basis of isolated pairs ac and bc, may be affected by the exchange terms between atoms a and b. An approximate calculation of this effect is possible on the basis of the model of the "caged" atom or molecule. In a system at high densities each atom or molecule is considered enclosed in a cage formed by the surrounding atoms. It is assumed that the effect of exchange terms between the caged atom and its nearest neighbors may be replaced by an ordinary potential term. For the evaluation of second order interactions this means that we may start from a zero order wave function which is a simple product of atomic wave functions:

$$\Psi_d = \Psi_{da} . \Psi_{db} . \Psi_{dc} \cdots \qquad (26)$$

and the Hamilton operator:

$$H = H_0 + H';$$

H' is the usual interaction operator.

The operator H_0 includes the potential term replacing the effect of exchange:

$$H_0 = \Sigma_a H_{oa}, \qquad (27)$$

where

$$H_{oa} = H_{oa} + V_a(r).$$

H_{oa} is the Hamiltonian for a free atom a, and $V_a(r)$ is the potential term

64

replacing exchange; r is written for the radius vectors of the electrons of atom a with the center of the atom as origin. The wave functions Ψ_{da}, Ψ_{db} ... are solutions of the equation:

$$H_{oa}\Psi_{da} = (E_{oa} + E_a')\,\Psi_{da}. \qquad (28)$$

E_{oa} is the energy of the ground state of a free atom and E_a' is the first order interaction of atom a with the surrounding atoms. E_a' is a function of the density of the system. Solutions of the wave equation (28) are possible for instance if $V_a(r)$ has the form of a boundary condition:

$$\begin{aligned} V_a(r) &= 0 && \text{for } r \leqslant R \\ &= \infty && \text{for } r > R; \end{aligned} \qquad (29)$$

R is called the radius of the cage of atom a. In principle, the value of R can be determined if E_{oa} and E_a' are known. In practice, one solves the wave equation for different values of R. Then it is assumed that the pressure

$$P = -\,(1/4\pi R^2)\,\partial E_a'/\partial R$$

which the electron "gas" of atom a exerts on the "wall" of the cage, balances the external pressure of the gas [28]). The wave equation (eq. 28) has been solved for hydrogen atoms [24] [25]), helium atoms [26]), the hydrogen molecule ion [27]) and argon atoms [28]).

With the new zero order wave functions (28) perturbation theory is applied up to the second order. The first order change in energy is identically zero, and in second order the dispersion forces are again additive, but now with respect to the new unperturbed state of caged atoms. If the second order interaction between a pair of caged atoms is written as $\Phi_{ij}^{(d)}$, where (d) refers to the dense state, then the total second order interaction is:

$$\Phi^{(d)} = \tfrac{1}{2}\Sigma_{i,j}'\,\Phi_{ij}^{(d)}, \qquad (30)$$

On the other hand the second order field as calculated on the assumption of additivity is:

$$\Phi^{(0)} = \tfrac{1}{2}\Sigma_{i,j}'\,\Phi_{ij}^{(0)} \qquad (30)$$

The superscript (o) refers to an isolated pair of atoms or molecules. If $\Phi^{(0)}$ is not equal to $\Phi^{(d)}$ then the dispersion forces are nonadditive with respect to isolated pairs. For the dipole part of the dispersion forces between two free atoms the expression is:

$$\Phi_{ij}^{(0)} = -\,3V^0\,(a^0)^2/4R_{ij}^6 \qquad (31)$$

V^0 is the first ionization potential of a free arom and a^0 is the polarizability of the free atom; R is the internuclear distance. In the same approximation the expression for a pair of caged atoms is:

$$\Phi_{ij}^{(d)} = -\,3V^d\,(a^d)^2/4R_{ij}^6. \qquad (31')$$

V^d and α^d are corresponding quantities for the caged atom in its unperturbed state. As follows from the calculations on the caged atom, both V^d and α^d are smaller than for the free atom. Physically this means that the effect of exchange terms is a compression of the electron cloud of the caged atom. As a result the intrinsic dipole moment of the atom decreases and the fundamental frequencies of the electrons increase. Therefore the dipole dispersion forces between a pair of caged atoms are smaller than for a pair of free atoms. A measure for this nonadditive effect will be given by the relative expression:

$$(\Phi_{ij}^{(d)} - \Phi_{ij}^{(0)})/\Phi_{ij}^{(0)} = 2\Delta a/a^0 + \Delta V/V^0, \qquad (32)$$

where we have substituted $a^d = a^0 + \Delta a$; $V^d = V^0 + \Delta V$; and only linear terms were taken into account. The values of $\Delta a/a^0$ and $\Delta V/V^0$ must be taken from the theoretical calculations on the caged atom or molecule; for argon from ref. 28. The formula (32) can be rewritten in terms of any set of intermolecular parameters; for the Lennard-Jones function this gives:

$$\Delta[\varepsilon\sigma^6] = (2\,\Delta a/a^0 + \Delta V/V^0)\,(\varepsilon\sigma^6)_0. \qquad (32')$$

The subscript (o) refers to the values of the intermolecular parameters as determined from low density data (second virial coefficients, transport properties). D e G r o o t and T e n S e l d a m [28] calculated for argon at a density of 600 Amagat and a temperature of 25°C the value $\Delta a/a^0 = -3.2 \times$ $\times 10^{-2}$. The value of $\Delta V/V^0$ is not known accurately; for helium it is always much smaller than the value of $\Delta a/a^0$ [26]. We take -6×10^{-2} as a lower limit for $(2\,\Delta a/a^0 + \Delta V/V^0)$ in the case of argon atoms.

For the approximate evaluation of the change in internal energy of the system it is further assumed that the distance R_{ij} between the centers of atoms i and j may be replaced by the distance between the centers of their cells. Then the additional internal energy, due to this nonadditive effect, is:

$$\Delta U = -(N/2)\,(2\,\Delta a/a^0 + \Delta V/V^0)\,4\,([\varepsilon\sigma^6]_0/R_0^6)\,\Sigma_{j=2}^N\,(R_0/R_{1j})^6, \qquad (33)$$

where R_0 is the distance between nearest neighbors in the lattice. At a density of 600 Amagat $R_0 = 4.3$ Å. For a face centered cubic array the lattice sum in (33) has the value 14.4; the intermolecular parameters ε and σ are 165.3×10^{-16} ergs and 3.405 Å, respectively as determined from the second virial coefficients. When these values are inserted into (33) the correction to the internal energy of the system is of the order of $+100$ cal per mole.

On account of the model used for the effect of exchange terms, this value is probably too high. The result indicates, however, that the use of a simple-product type of wave function for the evaluation of second order interactions in systems at high densities overestimates the magnitude of the dipole dispersion forces.

This model is not acccurate enough to give a quantitative comparison between nitrogen and carbon monoxide. It should be noted, however, that the ratio of the nonadditive effects, according to eq. (33), is proportional to $(\varepsilon_{CO}/\varepsilon_{N_2})(\sigma_{CO}/\sigma_{N_2})^6 = 1.16$. Also the Axilrod type of non-additive effect is more positive for carbon monoxide than for nitrogen. It must be expected that the non-additivity in the first order forces becomes more important at higher temperatures, because the interpenetration of charge distributions increases with increasing temperature.

§ 7. *Summary of results.* A comparison between the thermodynamic proper ties of nitrogen and carbon monoxide was based on a study of the intermole-cular field at high densities. The interaction field between two molecules was split into a spherical part (Lennard-Jones potential) and a part due to the anisotropie of the molecules (Φ_A). On the assumption that, as far as Φ_A is concerned, the molecules may be considered fixed at the centers of their cells at high densities, the thermodynamic functions are a sum of a "$L-J$" part and a "Δ" part. This makes it possible to apply the theorem of corresponding states to the $L-J$ parts of the thermodynamic functions of the two gases In this way we eliminate the differences due to the different values of the parameters ε and σ of the $L-J$ potential field, i.e. we compare the gases at the same "reduced" temperatures and densities. The rest of the calculation can then be based on a study of the potential field Φ_A. The dipole orientation and induction effects in carbon monoxide are too small to be of importance; the effect of the anisotropy in the London forces is, however, of considerable magnitude and must be taken into account. The theorem of corresponding states was based on the assumption that the parameters ε and σ are temper-ature and density independent; this condition is satisfied if the intermolecular forces are additive. It was shown, however, that a significant non additive effect in the attractive field between two molecules is caused by a deformation of electron clouds resulting from the overlap of wave functions of other molecules at high densities. An approximate expression was obtained for the change in the quantity $\varepsilon\sigma^6$; this change can be evaluated in a few simple cases by a correlation with theoretical calculations on caged atoms and mole-cules by several other authors. The result is that for argon the internal energy calculated with additive intermolecular forces may be too negative by an amount of 100 cal/mol. In this model the non additive effect is temper-ature independent.

So far, the permanent quadrupole moments of nitrogen and carbon mon-oxide have not been taken into account. This calculation and further conclusions which can be drawn from the properties of the solid states will be presented in a later publication.

REFERENCES

1) Michels, A., Wouters, H. and de Boer, J., Physica **1** (1934) 587; **3** (1936) 585, 597; Michels, A., Lunbeck, R. J. and Wolkers, G. J., Physica **17** (1951) 801.

2) W o u t e r s, H., Thesis, Amsterdam (1941).

3) M i c h e l s, A., L u p t o n, J. M., W a s s e n a a r, T. and d e G r a a f f, W., Physica **18** (1952) 121.

4) M i c h e l s, A., L u n b e c k, R. J. and W o l k e r s, G. J., Physica **18** (1952) 128.

5) W e n t o r f, R. H., B u e h l e r, R. J., H i r s c h f e l d e r, J. O. and C u r t i s s, C. F., J. chem. Phys. **18** (1950) 1484.

6) L e n n a r d-J o n e s, J. E. and D e v o n s h i r e, A. F., Proc. roy. Soc. **A 163** (1937) 53; **A 165** (1938) 1.

7) B e r l i n, T. H. and T h o m s e n, J. S., J. chem. Phys. **20** (1952) 1368.

8) L a x, M., J. chem. Phys. **20** (1952) 1351.

9) P o p l e, J. A., Proc. roy. Soc. **A 215** (1953) 67; Disc. Far. Soc. No. **15** (1953) 35.

10) v a n I t t e r b e e k, and d e C l i p p e l i e r, Physica **14** (1948) 349.

11) L e n n a r d-J o n e s, J. E. and I n g h a m, A. E., Proc. roy. Soc. **A 107** (1925) 636.

12) d e B o e r, J. H. and H e l l e r, G., Physica **4** (1937) 1045.

13) M a s s e y, H. S. W. and B u c k i n g h a m, R. A., Proc. roy. Irish. Acad. **A 45** (1938) 31.

14) d e B o e r, J., Physica **9** (1942) 363.

15) J a n s e n, L. and S l a w s k y, Z. I., J. chem. Phys. **22** (1954) 1701.

16) R o w l i n s o n, J. S. and C u r t i s s, C. F., J. chem. Phys. **19** (1951) 1519.

17) R o w l i n s o n, Disc. Far. Soc. No. **15** (1953) 52.

18) d e B o e r, J., Proc. roy. Soc. **A 215** (1952) 4.

19) L o n d o n, F.. Z. physik. Chem. **B 11** (1930) 222.

20) M a r g e n a u, H., Rev. mod. Phys. **11** (1939) 1.

21) A x i l r o d, B. M., and T e l l e r, E., J. chem. Phys. **11** (1943) 299.

22) A x i l r o d, B. M., J. chem. Phys. **17** (1949) 1349; **19** (1951) 719.

23) R o s e n, P h., J. chem. Phys. **21** (1953) 1007.

24) M i c h e l s, A., d e B o e r, J. and B i j l, A., Physica **4** *1937) 981; d e G r o o t, S. R. and t e n S e l d a m, C. A., Physica **12** (1946) 669; M i c h e l s, A. and d e G r o o t, S. R., Physica **16** (1950) 183.

25) S o m m e r f e l d, A. and W e l k e r, H., Ann. Physik **32** (1938) 56.

26) t e n S e l d a m, C. A. and d e G r o o t, S. R., Physica **18** (1952) 891; **18** (1952) 905.

27) C o t t r e l l, T. L., Trans. Far. Soc. **47** (1951) 337.

28) t e n S e l d a m, C. A. and d e G r o o t, S. R., Physica **18** (1952) 910; t e n S e l d a m, C. A., Thesis, Utrecht (1953).

PART II. QUADRUPOLE MOMENTS AND PROPERTIES OF THE SOLID STATES

§ 1. *Introduction.* In a previous section [1]) the thermodynamic properties of compressed nitrogen and carbon monoxide as determined from experimental p-v-T-data were compared in the density region between 400 and 600 Amagat and at temperatures between 0°C and 150°C. The analysis was based on the assumption that the intermolecular field of nitrogen and carbon monoxide may be written as the sum of a Lennard-Jones potential field, a part due to multiple orientation and a correction term for the nonadditive part of the intermolecular field at high densities. On the assumption that, as far as the orientational and nonadditive parts of the potential field are concerned, the molecules may be considered fixed at the centers of their cells at high densities, the thermodynamic functions are additive in the respective terms of the potential field. This enables us to evaluate the differences in thermodynamic functions between nitrogen and carbon monoxide due to the differences in the orientational and nonadditive parts by an application of the

68

theorem of corresponding states. It was found that the effect of anisotropy in the dipole dispersion forces may not be neglected at high densities. Further a significant nonadditive effect exists in the London forces between two molecules, due to the overlap of wave functions of the nearest neighbours. The effect of dipole orientation and induction of the permanent dipoles of carbon monoxide is negligible. So far, the permanent quadrupole moments of the two molecules have not been taken into account. This part of the analysis deals with the calculation of the values for the quadrupole moments with the help of experimental data on the sublimation energies of the crystals, extrapolated to 0°K.

§ 2. *The crystal structures of nitrogen and carbon monoxide.* Nitrogen and carbon monoxide crystallize in two alltropic modifications, called α and β. The α-form is face centered cubic and stable at the lowest temperatures (space group $T^4 - P2_13$). The β-form has hexagonal symmetry. The transition temperature for $\alpha - \beta$ N_2 is 35.5°K, for $\alpha - \beta$ CO 61.5°K.

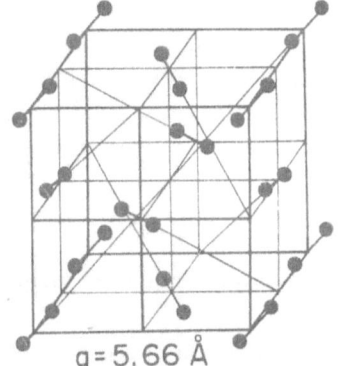

a = 5.66 Å

Fig. 1. Crystal structure $\alpha - N_2$.

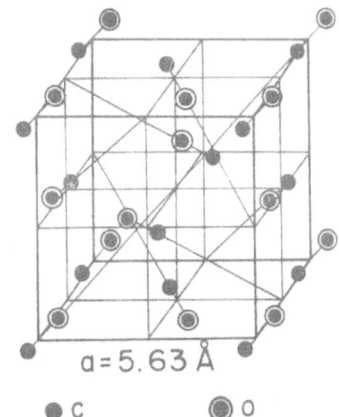

a = 5.63 Å

● C   ◎ O

Fig. 2. Crystal structure of $\alpha - CO$.

Determination of the crystal structures was performed from X-ray analyses by V e g a r d [2] [3] [4] [5]) and R u h e m a n n [6]). We will consider only the structure of the α-forms. It appears that the α-forms for N_2 and CO are almost identical; the same is true for the β-modifications. Fig. 1 shows the structure of α-N_2, fig. 2 that of α-CO. The distance between nearest neighbors in α-N_2 is 4.00 Å and in α-CO 3.98 Å at 20°K. The axes of the molecules in the α-modifcations are directed towards the body centers of the cubes. From the X-ray analyses of the β-structures it appears that there is considerable scattering around the equilibrium orientations. (V e g a r d, *loc. cit.*). The density of α-CO crystals is 1.0288 gr/cm³ at 20°K, of α-N_2 1.0265 gr/cm³ at the same temperature. The density of β-CO ,however, is less than that of β-N_2; the values are 0.929 and 0.982 at 65°K and 63°K, respectively.

K e l l e y [7]) gives the following values for the sublimation energies of the crystals, extrapolated to absolute zero:

α-N$_2$ 1652 cal/mol;   α-CO   1905 cal/mol.

§ 3. *The crystal at absolute zero.* In classical statistics the energy of a molecular crystal at absolute zero is given by the sum of interactions between isolated pairs of molecules, if additivity of the intermolecular forces is assumed. For a face centered cubic lattice, using the lattice sums given by L e n n a r d-J o n e s and I n g h a m [8]), and a Lennard-Jones potential field, one obtains:

$$\varphi^* = 6.066/V^{*4} - 14.454/V^{*2}, \tag{1}$$

where $\varphi^* = \varphi/N\varepsilon$ and $V^* = V/N\sigma^3$; V is the volume per mol. This equation gives for the reduced potential energy and volume at absolute zero:

$$\varphi_0^* = -8.610 \text{ and } V_0^* = 0.916. \tag{2}$$

The density of the crystals of nitrogen and carbon monoxide, calculated from (2) are $d_{N_2} = 0.978$ and $d_{CO} = 0.952$ at absolute zero. These values should be compared with 1.0265 and 1.0288 at 20°K which are the experimental values for nitrogen and carbon monoxide, respectively. The theoretical values are lower than the experimental data; moreover the difference between nitrogen and carbon monoxide has the wrong sign.

Deviations from the classical values (2) can be explained on the following basis:

a) the zero-point energy of the crystal must be added to the classical potential energy at 0°K;

b) orientation and induction forces between the molecules must be taken into the account;

c) the Lennard-Jones potential field is not accurate enough for the evaluation of properties of the solid states;

d) the deviations are caused primarily by nonadditive effects in the potential field at high densities.

For the *zero-point energy* of the crystal we use the Debije expression:

$$U_{z.p.} = 9 Nk\theta/8,$$

where θ is the Debije characteristic temperature. Calculations of θ have been performed by H e r z f e l d and G o e p p e r t-Ma y e r [10]), K a n e [11]), D e B o e r and B l a i s s e [9]), C o r n e r [12]) and D e B o e r and L u n-b e c k [13]). Following Corner's method we have, for a face centered cubic lattice and an additive Lennard-Jones potential field:

$$U_{z.p.}^* = 9\,\theta^*/8 = -\frac{0.2913}{V^{*1/3}}\, \Lambda^* \left[\frac{265.30}{V^{*4}} - \frac{128.02}{V^{*2}} \right]^{1/2}, \tag{3}$$

70

where $\Lambda^* = h/\sigma \, (m\varepsilon)^{1/2}$. The internal energy at absolute zero for a Lennard-Jones type of potential field is then:

$$U_0^* = \varphi_0^* + 9\,\theta_0^*/8 \tag{4}$$

as a function of V^* and Λ^*. With the condition $P_0^* = -\,\partial U_0^*/\partial V^* = 0$ the reduced volume at absolute zero can be evaluated and thus also U_0^*. In the following table theoretical and experimental values of V_0^*, $-U_0^*$ and θ_0^* are compared for crystals of argon, nitrogen and carbon monoxide. The experimental values of U_0^* were obtained from the sublimation energies (K e l l e y ref. 7) and expressed in reduced form (see also Lunbeck, ref. 13). In the last column values of $U_{0\mathrm{exp.}} - U_{0\mathrm{th.}}$ are listed for the three crystals. The values of θ_0 were obtained using the experimental values of V_0 for the best accuracy.

TABLE I

	Properties of argon, nitrogen and carbon monoxide crystals at absolute zero.						
	V_0^*		θ_0^*		$-U_0^*$		$(U_0 - U_0)$ exp. th. cal/mol
	th.	exp.	th.	exp.	th.	exp.	
classical	0.916				8.61		
argon	0.956	1.00	0.66	0.67	7.63	7.77	— 33
nitrogen	0.962	0.873	0.78	0.69	7.64	8.63	—190
carbonmonoxide	0.957	0.848	0.77	—	7.50	9.52	—402

Comparison between the three crystals shows that the potential energy at absolute zero, corrected for the zero-point energy ,is somewhat too positive for argon, but much too positive for nitrogen and carbon monoxide, if the calculations are based on a Lennard-Jones potential field and additivity of the intermolecular forces is assumed.

The deviations for the rare gases can be interpreted as an inaccuracy of the Lennard-Jones potential field if the validity of additivity of the intermolecular forces is retained (K i h a r a, ref. 16). Although this empirical procedure of correcting the Lennard-Jones potential at high densities is the most useful method for practical applications it can be shown that deviations from the principle of additivity do occur and are significant in the solid states (A x i l r o d [17]), R o s e n [18]); also part I of this chapter). Since no accurate theory is available we will correct the values for N_2 and CO by about -35 cal/mol, i.e. of the same order of magnitude as for argon.

The calculation of the characteristic temperature θ was based on an additive Lennard-Jones potential field. Non additive effects will alter θ slightly; this change in the characteristic temperature will be neglected for the following calculations. The deviations for nitrogen and carbon monoxide, corrected for what may be called the "high density effect", discussed above, must be due to strong orientation forces. It will be assumed that the effect of the orientation and induction forces on the characteristic temperature may be neglected. The deviation for a-CO is more than twice that of a-N$_2$. In the β-forms the molecules have considerable rotational freedom. This may explain that the transition temperature a-β of CO lies higher than that of N$_2$ and also that the difference in densities between nitrogen and carbon monoxide changes sign going from the a to the β crystals.

§ 4. *Orientation and induction effects at* $0°K$. First we consider the interaction between the permanent and induced multipoles of nitrogen and carbon monoxide. The orientation interaction between two identical molecules with a permanent dipole moment μ and a permanent quadrupole moment Q, directed along the length axis of the molecule, is given by (see fig. 3):

$$V_{or} = - (\mu^2/R^3) [2 \cos \theta_1 \cos \theta_2 - \sin \theta_1 \sin \theta_2 \cos (\varphi_1 - \varphi_2)] +$$
$$+ (3/2R^4) \mu Q \{[\cos \theta_2 + 2 \cos \theta_1 \sin \theta_1 \sin \theta_2 \cos (\varphi_1 - \varphi_2) - 3 \cos^2 \theta_1 \cos \theta_2] -$$
$$- [\cos \theta_1 + 2 \cos \theta_2 \sin \theta_2 \sin \theta_1 \cos (\varphi_1 - \varphi_2) - 3 \cos^2 \theta_2 \cos \theta_1]\} +$$
$$+ (3/4R^5) Q^2 \{1 - 5 \cos^2 \theta_1 - 5 \cos \theta_2 - 15 \cos^2 \theta_1 \cos^2 \theta_2 +$$
$$+ 2 [4 \cos \theta_1 \cos \theta_2 - \sin \theta_1 \sin \theta_2 \cos (\varphi_1 - \varphi_2)]^2\}, \qquad (5)$$

if the quadrupole moment is defined by:

$$Q = \Sigma_i \, e_i \, (z_i^2 - x_i^2) \qquad (6)$$

and z_i is measured along the length axis of the axially symmetric molecule. Due to the symmetry of the crystal the dipole-quadrupole orientation effect is identically zero. The first term in (5) represents the dipole-dipole, the last term the quadrupole-quadrupole orientation. For the summation over the crystal lattice we take one molecule as a center and sum the contributions due to 12 nearest neighbours at distance r_0, 6 next nearest neighbours at distance $r_0\sqrt{2}$, in the third shell 24 molecules at a distance $r_0\sqrt{3}$ and 12 molecules at distance $2r_0$. The following table gives the dipole-dipole and quadrupole orientation effects due to 54 molecules around the molecule at the center.

TABLE II

Shell No.	distance from center	number of molecules	orientation interaction for each pair	
			dipole μ^2/r_0^3	quadrupole Q^2/r_0^5
1	r_0	12	$-1/3$	$-19/12$
2	$r_0\sqrt{2}$	6	0	$-7\sqrt{2}/24$
3	$r_0\sqrt{3}$	12	$+\sqrt{3}/9$	$-31\sqrt{3}/2916$
	$r_0\sqrt{3}$	12	$-\sqrt{3}/27$	$+\sqrt{3}/324$
4	$2r_0$	6	$+ 1/8$	$+ 9/128$
	$2r_0$	6	$- 1/8$	$- 13/384$

This gives for the total dipole and quadrupole orientation energy:
$$V_{dip} = (N/2) (- 2.46) \, \mu^2/r_0^3 = - 3.4 \text{ cal/mol}, \text{ with } \mu = 0.11 \text{ D and}$$
$r_0 = 3.98$ Å for carbon monoxide.

$V_{qu} = (N/2)\,(-21.413)$. $Q^2/r_0^5 = -154.7\,q^2$ cal/mol for carbon monoxide and $-150.9\,q^2$ cal/mol for nitrogen; q is a number which measures the quadrupole moment Q in units 10^{-26} e.s.u.

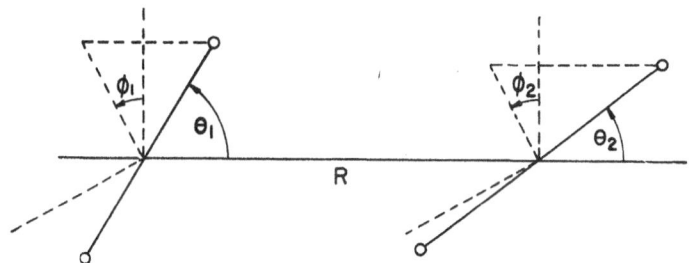

Fig. 3. Angular coordinates of interacting axial molecules.

For the evaluation of the *induction effects* we consider a molecule at the center of a coordinate system (x, y, z). The z-axis is directed along the length axis of the molecule. (see fig. 4).

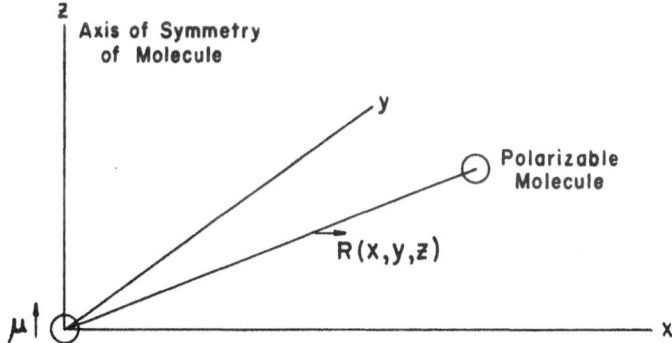

Fig. 4. Choice of the coordinate system for two interacting molecules.

The potential at a distance $\mathbf{R}(X, Y, Z)$ from the origin, due to the dipole and quadrupole moment of the central molecule, is:

$$\varphi_{ind.} = (1/R^3)\,(\boldsymbol{\mu}.\mathbf{R}) - (1/2R^5)\,(\mathbf{M}.\mathbf{Q}), \tag{7}$$

where the components of $\mathbf{M}$ and $\mathbf{Q}$ are given by: $M_x = R^2 - 3X^2$, etc.; $Q_x = \Sigma_i\,e_i\,x_i^2$, etc.

The field $\mathbf{F}$ is:

$$\mathbf{F} = -\nabla\varphi_{ind.} = -\boldsymbol{\mu}/R^3 + (3/R^5)\,\mathbf{R}(\boldsymbol{\mu}.\mathbf{R}) + \mathbf{N}/R^5 - (5/2R^7)\,\mathbf{M}(\mathbf{Q}.\mathbf{R}), \tag{8}$$

where $\mathbf{N}$ is a vector with components $N_x = (Q_y + Q_z - 2Q_x)\,.X$, etc. N_y and N_z are formed by cyclical permutation of x, y and z. It will be assumed that the polarizability of the molecule at $\mathbf{R}$ is isotropic, i.e. we use for nitrogen and carbon monoxide $a = \bar{\alpha} = (a_x + a_y + a_z)/3$. In the crystal we take the molecules at **R** as the center and sum again over 54 nearest molecules. The induced energy at the central molecule, V_{ind}, is:

$$V_{\text{ind}} = - (a/2) \cdot [\Sigma_j \, \mathbf{F}_j]^2, \tag{9}$$

where $\mathbf{F}_j$ is the field induced at the central molecule by molecule j. The symmetry of the crystals causes several induction effects to be identically zero. These are:

1) quadrupole-induced quadrupole interaction;

2) interaction between the induced dipole/quadrupole at the center and the permanent quadrupoles/dipoles of the surrounding molecules.

The only induction effect which is not zero is the interaction between the induced dipole at the center and the permanent dipoles of the other molecules Table III gives the field **F** at the central molecule due to four shells of surrounding molecules.

TABLE III

	Induced dipole effect and anisotropy in London forces due to 54 nearest molecules			
Shell	distance from center	number of molecules	induced dipole effect field **F** for each shell $. \mu/r_0{}^3)$	anisotropy effect in London dispersion forces for each pair $. \epsilon(\sigma/r_0)^6$
1	r_0	12	$+ 4\sqrt{3}/3 \cdot e_F$	$+ (10/3) \, \gamma^2$
2	$r_0\sqrt{2}$	6	0	$+ (1/2) \, \gamma^2$
3	$r_0\sqrt{3}$	12	$- 4/9 \cdot e_F$	$- (8/81) \, \gamma \; + (2/81) \, \gamma^2$
	$r_0\sqrt{3}$	12	$- 4/9 \cdot e_F$	$+ (8/81) \, \gamma \; + (2/81) \, \gamma^2$
4	$2r_0$	6	0	$+ (1/16) \, \gamma \; - (3/32) \, \gamma^2$
	$2r_0$	6	0	$- (1/16) \, \gamma \; + (1/32) \, \gamma^2$

$\mathbf{e}_F$ is a vector with components $+1, +1, -1$ along the cubic axes.
The total induced dipole energy is, per mol:

$$V_{ind} = - a \, (N/2) \, (6.053) \, \mu^2/r_0^6 = - 0.24 \; \text{cal/mol},$$

with $a = 1.844 \times 10^{-24}$ cm^3 for carbon monoxide [19]).

§ 5. *Anisotropy effect in London dispersion forces.* If the charge distribution in a molecule is not spherically symmetric, the dispersion forces are anisotropic. The correction term in the induced dipole field between two molecules is (see ref. 1):

$$\varphi_{\text{an}} = 4\varepsilon(\sigma/R)^6 \, [(1 - \tfrac{3}{2}\cos^2\theta_1 - \tfrac{3}{2}\cos^2\theta_2) \, \gamma +$$
$$+ \{\tfrac{3}{2}\cos^2\theta_1 + \tfrac{3}{2}\cos^2\theta_2 - \tfrac{3}{2}(2\cos\theta_1\cos\theta_2 - \sin\theta_1\sin\theta_2\cos(\varphi_1 - \varphi_2))^2\} \, \gamma^2], \tag{10}$$

where the anisotropy factor γ is defined by:

$$\gamma = (a_z - a_y)/3\bar{a}; \quad \bar{a} = (a_z + 2a_y)/3$$

and the z-axis is taken along the length axis of the molecule; $a_x = a_y$ for axially symmetric molecules. The eq. (10) was derived on the assumption that the fundamental vibrational frequencies of the molecule along and perpendicular to the molecular axis are the same. Again we sum the contributions of the 54 nearest molecules around a central molecule in the crystal lattice. The results are given in table III, last column. Terms linear in γ cancel, as is seen from the table. For the total anisotropy effect per mol we obtain:

$$V_{an} = (N/2)\,(43.218)\,\varepsilon(\sigma/r_0)^6\,\gamma^2 = +\,86.1 \text{ cal/mol}$$

for carbon monoxide and $+88.4$ cal/mol for nitrogen[*]).

The values of ε and σ were obtained from experimental second virial coefficients (see ref. 1 and ref. 13). Therefore, nonadditive terms in the intermolecular field have been neglected for the calculation of the anisotropy effect. Further, no correction was made for the change in polarizability with the density of the system.

§ 6. *Evaluation of the quadrupole moments.* With the calculations of the preceding sections the evaluation of the quadrupole moments can now be performed. Since the dipole-dipole orientation and induction effects in CO are very small, we neglect these contributions to the internal energy of the crystal. Using the values of $U_{0\,exp} - U_{0\,th}$ of table I, and including corrections for the anisotropy in the London forces, and the approximate values for the "high density" effect, we have:

nitrogen: $243 = 150.7\,q^2$; $Q = q \times 10^{-26}$ e.s.u. $= 1.27 \times 10^{-26}$ e.s.u.

carbon monoxide: $453 = 154.7\,q^2$; $Q = q \times 10^{-26}$ e.s.u. $= 1.71 \times 10^{-26}$ e.s.u.

Due to the uncertainty in the magnitude of the "high density" effect, these are only approximate values. If this effect is not taken into account, then the values of the quadrupole moments are lower by about seven percent.

In recent years S m i t h and H o w a r d [20]) and H i l l and S m i t h[21]) reported investigations of the broadening of the 3-3 inversion line of NH_3 by other gases. From the data obtained they calculated a collision diameter for the NH_3-foreign gas collision. In several cases this diameter was markedly larger than that given by kinetic theory. The authors ascribed the broadening effect in these cases to the interaction of the dipole moment of NH_3 and a permanent quadrupole moment of the foreign gas, averaged over the rotation. The quadrupole moment of the foreign gas molecule was calculated in this way. For the quadrupole moment of nitrogen, S m i t h and H o w a r d obtain a value of 1.29×10^{-26} e.s.u.; for carbon monoxide, H i l l and S m i t h calculate 1.62×10^{-26} e.s.u.

A n d e r s o n [22]) has shown that the interaction of the quadrupole

---

[*]).$\gamma_{CO}=0.167$; $\gamma_{N_2}=0.176$ as given by L a n d o l t B o r n s t e i n. K. G. D e n b i g h, Trans. Far. Soc. **36** (1930) 936, lists $\gamma_{N_2} = 0.189$ and $\gamma_{CO} = 0.168$. In the calculation above we used a mean value 0.182 for nitrogen.

moment of NH_3, calculated from its structure, and the dipole induced in the foreign gas molecule gives a collision diameter in good agreement with those molecules whose experimental collision cross sections approach kinetic theory values.

For nitrogen, carbon monoxide and other molecules, however, this diameter is too small, and the assignment of a permanent electric quadrupole to these molecules seems justified.

The values determined from the sublimation energies are seen to agree well with those calculated from collision diameters.

Just recently P o p l e [23]) calculated the intermolecular parameters of carbon dioxide from an analysis of experimental second virial coefficients and the sublimation energy of the crystal. The intermolecular field was assumed to be of the Lennard-Jones form, plus a quadrupole orientation effect. He obtained excellent agreement with the experimental data for a value of the quadrupole moment of 5.73×10^{-26} e.s.u., compared with 3.12×10^{-26} as determined from micro wave spectra [20]). Since the crystal of CO_2 has the same symmetry properties as the α-forms of nitrogen and carbon monoxide, the anisotropy effect in the London forces for carbon dioxide increases the value of the quadrupole moment.

REFERENCES

1) Part I, Chapter V of this thesis.
2) V e g a r d, L., Z. Phys. **61** (1930) 185.
3) V e g a r d, L., Z. Phys. **88** (1934) 235.
4) V e g a r d, L., Z. Phys. **58** (1929) 497.
5) V e g a r d, L., Z. Phys. **79** (1932) 471.
6) R u h e m a n n, M., Z. Phys. **76** (1932) 368.
7) K e l l e y, K. K., Bulletin **383**, U.S. Dept. of the Interior, Bur. of Mines (1935).
8) L e n n a r d J o n e s, J. E. and I n g h a m, A. E., Proc. roy. Soc. **A 107** (1925) 636.
9) d e B o e r, J. and B l a i s s e, B. S., Physica **14** (1948) 149.
10) H e r z f e l d, K. F. and G o e p p e r t-M a y e r, M., Phys. Rev. **46** (1934) 995.
11) K a n e, G., J. chem. Phys. **7** (1939) 603.
12) C o r n e r, J., Trans. Far. Soc. **35** (1939) 711.
13) L u n b e c k, R. J., Thesis, Amsterdam (1951).
14) W e n t o r f, R. H., B u e h l e r, R. J., H i r s c h f e l d e r, J. O. and C u r t i s, C. F., J. chem. Phys. **18** (1950) 1484.
15) K i h a r a, T., Rev. mod. Phys. **25** (1953) 831.
16) A x i l r o d, B. M., J. chem. Phys. **17** (1949) 1349; **19** (1951) 719.
17) R o s e n, P h., J. chem. Phys. **21** (1953) 1007.
18) M a r g e n a u, H., Rev. mod. Phys. **11** (1939) 1.
19) v a n I t t e r b e e k, A. and d e C l i p p e l i e r, K., Physica **14** (1948) 349.
20) S m i t h, W. V. and H o w a r d, R., Phys. Rev. **79** (1950) 132.
21) H i l l, R. M. and S m i t h, W. V., Phys. Rev. **82** (1951) 451.
22) A n d e r s o n, P. W., Phys. Rev. **80** (1950) 511.
23) P o p l e, J. A., Proc. roy. Soc. **A 221** (1954) 508.

PART III. QUADRUPOLE ORIENTATION AND INDUCTION EFFECTS AT HIGH TEMPERATURES

§ 1. *Introduction.* In two previous sections [1] [2] (herafter referred to as I and II, respectively) the intermolecular field of compressed nitrogen and carbon monoxide was analyzed on the basis of experimental pvT-data. The interaction field between two molecules was written as a spherically symmetric part, satisfying the conditions for the validity of the theorem of corresponding states, and a part due to orientation and induction effects. The spherically symmetric part has the general twoparameter form:

$$\Phi_{ij} = \varepsilon . f(\sigma/r_{ij}) \tag{1}$$

where r_{ij} is the distance between the centers of molecules i and j. It is assumed that the Lennard-Jones 12 : 6-function gives a representation of Φ_{ij} which is accurate enough for the present purposes, as far as the values of the parameters ε and σ are concerned. Numerically, these parameters are determined from low density pvT-data (second virial coefficients). In this formulation deviations at high densities from the values of the thermo-dynamic properties given by the theorem of corresponding states must be analyzed on the basis of three possibilities:

a) deviations expected also for atoms or molecules with a spherically symmetric potential field (nonadditive effects); argon was taken as a standard of comparison.

b) those expected for nitrogen and carbon monoxide, but not for argon (orientation and induction effects of permanent multipoles higher than dipoles; anisotropy in the dispersion forces).

c) deviations which must be attributed to carbon monoxide alone (dipole orientation and induction effects).

The application of the theorem of corresponding states to the comparison between nitrogen and carbon monoxide is possible on the following approxmation: At high densities (between 400 and 600 Amagat) the molecules may be considered fixed at the centers of their cells, as far as the orientation, induction and nonadditive parts of the potential field are concerned. The differences in thermodynamic properties between the two gases, due to the different values of the parameters ε and σ, can then be eliminated by comparing the two gases at the same "reduced" temperatures and densities. On the basis of the free volume theory of L e n n a r d-J o n e s and D e-v o n s h i r e, H i r s c h f e l d e r c.s.[3] computed thermodynamic properties of nitrogen up to a density of 480 Amagat. The calculated internal energy at 480 Amagat and a temperature of 0°C is −640 cal/mol, whereas the experimental value is −695 cal/mol. For argon at 640 Amagat and in the same temperature region the calculated internal energy is too negative by about ten percent. The deviations for argon must be explained in terms of statistical inaccuracies of the free volume theory, errors due to the approximate nature of the Lennard-Jones potential and to deviations from additivity of the intermolecular field. The calculations for nitrogen were based on a Lennard-Jones 12 : 6-potential. Taking argon as a standard of comparison it can be concluded that the contributions to the internal energy for nitrogen, due to orientation and induction effects, are of the order of twenty percent at 480 Amagat.

Evaluations of the dipole orientation and induction effects in carbon monoxide and of the effect of anisotropy in the dispersion forces for nitrogen and carbon monoxide were performed at high densities (I. § 4,5). As a test for the theoretical approach followed in this analysis and to obtain independent information, the values of the quadrupole moments were determined from experimental sublimation energies of the crystals, extrapolated to 0°K. The result is: (II)

$$Q_{N_2} = 1.27 \times 10^{-26} \text{ e.s.u.}; \quad Q_{CO} = 1.71 \times 10^{-26} \text{ e.s.u.}$$

These values agree well with those determined from micro wave spectra [4]. (1.29×10^{-26} e.s.u. and 1.62×10^{-26} e.s.u., respectively).

With the help of these values, the theoretical analysis can now be concluded by calculating the quadrupole orientation interaction, quadrupole induction effect and the dipole-quadrupole orientation (in carbon monoxide) at densities between 400 and 600 Amagat and for temperatures between 0°C and 150°C.

§ 2. *Quadrupole orientation interaction.* The partition function for the quadrupole orientation effect is, with the approximation outlined above:

$$Z_N^{(qu)} = \int \ldots \int e^{-\Phi(qu)/kT} \, d\omega_N, \tag{2}$$

where $\Phi^{(qu)} = \frac{1}{2} \Sigma'_{i,j} \Phi_{ij}^{(qu)}$

$$\Phi_{ij}^{(qu)} = (3Q^2/4R_{ij}^5) [1 - 5\cos^2\theta_i - 5\cos^2\theta_j - 15\cos^2\theta_i\cos^2\theta_j +$$
$$+ 2\{4\cos\theta_i\cos\theta_j - \sin\theta_i\sin\theta_j\cos(\varphi_i - \varphi_j)\}^2]; \tag{3}$$

$\int d\omega_N$ has been written for integration over the orientations of the molecules; θ and φ are the angular coordinates of a molecule.

Q is the quadrupole moment and R_{ij} is the distance between the centers of cells i and j. The integrand of (2) is expanded in a series of powers of $\Phi^{(qu)}/kT$:

$$Z_N^{(qu)} = \Sigma_{p=0}^{\infty} (-1)^p/p! \cdot \int \ldots \int [\Phi^{(qu)}/kT]^p \, d\omega_N = (4\pi)^N (1 + \Sigma_{p=1}^{\infty} M_p),$$

with

$$M_p = (4\pi)^{-N} \cdot (-1)^p/p! \cdot \int \ldots \int [\Phi^{(qu)}/kT]^p \, d\omega_N. \tag{4}$$

The term M_1 gives the approximation in which each quadrupole is

78

considered as independent of the others; therefore M_1 is zero. The term M_2 becomes:

$$M_2 = (4\pi)^{-N} N \int \ldots \int \{\Sigma'_{j \neq k} \Phi^{(qu)}_{1j} \Phi^{(qu)}_{1k} + \tfrac{1}{2} \Sigma'_j \Phi^{(qu) \, 2}_{1j}\} \, d\omega_N/2(kT)^2.$$

The prime in the summation indicates that the sum is expended over all molecules except number 1. Integration over the orientations has the result:

$$(4\pi)^{-N} \int \ldots \int \Phi^{(qu)}_{1j} \Phi^{(qu)}_{1k} \, d\omega_N = 0$$

$$(4\pi)^{-N} \int \ldots \int \Phi^{(qu)2}_{1j} \, d\omega_N = + 14 \, Q^4/5R^{10}_{1j}.$$

The expression for the partition function is in this approximation:

$$\ln Z^{(qu)}_N = (4\pi)^{-N} N \Sigma'_j \int \ldots \int \Phi^{(qu)2}_{1j} \, d\omega_N/4(kT)^2$$

and the contribution to the internal energy of the system is:

$$U^{(qu)}_N = - \partial \ln Z^{(qu)}_N/\partial \, 1/kT = - (7RT/5) \, [Q^4/(kT)^2 \, R^{10}_0] . \Sigma'_j \, (R_0/R_{1j})^{10}.$$

R_0 is the distance between nearest neighbors in the lattice. The value of the lattice sum for a face centered cubic array has been evaluated by L e n- n a r d-J o n e s and I n g h a m [5]) to be 12,3112. If a hexagonal closest packing is taken as the best approximation, then the lattice sum is 12.3124 [6]).

At a density of 600 Amagat $R_0 = 4.3$ Å; an average value of Q for carbon monoxide and nitrogen is 1.5×10^{-26} e.s.u. Inserting these values into the expression for the internal energy gives:

$$U^{(qu)}_N = - 14.3 \text{ cal/mol for } Q = 1.5 \times 10^{-26} \text{ e.s.u. and } T = 300°\text{K}.$$

The result shows that the quadrupole orientation energy at temperatures between 0°C and 150°C is too small to contribute significantly to the molecular interaction in compressed nitrogen and carbon monoxide up to densities of 600 Amagat.

The same formalism can be used to include the orientation effect between the permanent dipoles and quadrupoles of carbon monoxide, with analogous eqs. (2, 3, 4) and $\Phi^{(qu)}$ replaced by $\Phi^{(qu)} + \Phi^{(d-qu)}$. Cross terms in M_2 are zero on integration. Comparison shows that the dipole-quadrupole orientation energy is only about one fifth of the quadrupole orientation energy (dipole moment of CO is 0.1172 D [7]). Higher terms in the series expansion for $\ln Z_N$ need not be evaluated.

§ 3. *Quadrupole induction effect.* In I. §4b the induction effect of the permanent dipoles of carbon monoxide at high temperatures was calculated. In this section the evaluation will be extended to molecules having permanent quadrupole moments. The notation is the same as used for the evaluation of induced energy in the crystals at absolute zero. (II. eqs. 7, 8 and fig. 4).

Consider a molecule with a permanent dipole moment μ and a permanent quadrupole moment Q at the origin of a coordinate system (x, y, z). The potential at a point $\mathbf{R}(X, Y, Z)$ is [8]:

$$\Phi_{\text{ind}} = (1/R^3)\,(\boldsymbol{\mu}.\mathbf{R}) - (1/2R^5)\,(\mathbf{M}.\mathbf{Q}), \qquad (5)$$

where the components of $\mathbf{M}$ and $\mathbf{Q}$ are given by: $M_x = R^2 - 3X^2$, etc.; $Q_x = \Sigma_i\,e_i\,x_i^2$, etc. The field strength at $\mathbf{R}$ is:

$$\mathbf{F} = -\,\boldsymbol{\nabla}\Phi_{\text{ind}} = -\,\boldsymbol{\mu}/R^3 + (3/R^5)\,\mathbf{R}(\boldsymbol{\mu}.\mathbf{R}) + \mathbf{N}/R^5 - (5/2R^7)\,\mathbf{M}(\mathbf{Q}.\mathbf{R}); \quad (6)$$

$\mathbf{N}$ is a vector with components $N_x = (Q_y + Q_z - 2Q_x)\,X$; N_y and N_z are formed by cyclical permutation of x, y and z. It is assumed that the polarizability of the molecule at $\mathbf{R}$ is isotropic, i.e. we use for nitrogen and carbon monoxide $a = \bar{a} = (a_x + a_y + a_z)/3$.

The induced energy at $\mathbf{R}$, due to all the other molecules, is:

$$V_{\text{ind}}^{(1)} = -\,(a/2)\,[\Sigma'_j\,\mathbf{F}_j^{(1)}]^2$$

$\mathbf{F}_j^{(1)}$ is the field at the center of molecule 1 (at $\mathbf{R}$) due to molecule j. The total induced energy is:

$$V_{\text{ind}} = \Sigma_{i=1}^N\,V_{\text{ind}}^{(i)}$$

and the partition function becomes:

$$Z_N^{(\text{ind})} = \int \ldots \int e^{-V_{\text{ind}}/kT}\,\mathrm{d}\omega_N. \qquad (7)$$

As in the previous calculations the molecules are considered as located at the centers of their cells. On expanding (7) the first term is:

$$M_1 = (Na/2kT)\,(4\pi)^{-N}\int \ldots \int (\Sigma_{j=2}^N\,\mathbf{F}_j^{(1)})^2\,\mathrm{d}\omega_N.$$

Integration has the result:

$$M_1 = (Na/kT)\,\{(\mu^2/R_0^6).\Sigma_{j=2}^N\,(R_0/R_{1j})^6 + (3Q^2/2R_0^8).\Sigma_{j=2}^N\,(R_0/R_{1j})^8\} \qquad (8)$$

The dipole induced energy was already evaluated in I. § 4b, therefore we restrict ourselves to the second term in (8).

As average values we take $a = 1.84 \times 10^{-24}$ cm^3; $Q = 1.5 \times 10^{-26}$ e.s.u. for nitrogen and carbon monoxide. The lattice sum $\Sigma'_j\,(R_0/R_{1j})^8$ is equal to 12.8019 for a face centered cubic array (L e n n a r d-J o n e s and I n g-h a m, loc. cit.). Then the value of the quadrupole induced energy is, at a density of 600 Amagat:

$$U_N^{(\text{ind})} = -\,\partial \ln Z_N^{(\text{ind})}/\partial\,1/kT = -\,9.9 \text{ cal/mol}.$$

It should be noted that the induced energy due to the quadrupole moments is of the same order of magnitude as the quadrupole orientation energy at 600 Amagat and 300°K. However, both effects are small compared with the orientation effect due to anisotropy in the dispersion forces for temperatures between 0°C and 150°C (I. § 5).

§ 4. *Summary of results.* The results of this analysis of molecular inter-action in compressed nitrogen and carbon monoxide may be summarized as follows:

a) In the low density region (second virial coefficients) and for temper-atures between 0°C and 150°C the potential field can be adequately described with the Lennard-Jones 12 : 6-function. The theorem of corresponding states is valid in this region with high accuracy.

b) At high densities (between 400 and 600 Amagat) and for temperatures between 0°C and 150°C the simple representation of the intermolecular field by a spherically symmetric function is in error by as much as twenty percent for the internal energy. The most important deviation is caused by the orientation effect due to anisotropy in the dispersion forces. The corresponding contribution to the internal specific heat at constant volume is of the order of 0.5 cal/mol. degree at 600 Amagat and 300°K. The effect of quadrupole orien-tation and induction is small compared with the anisotropy contribution in this region.

c) In the crystals of nitrogen and carbon monoxide below the $\alpha - \beta$ transition temperatures, the quadrupole orientation forces constitute the most important deviation from spherical symmetry of the interaction field. Orientation effects due to anisotropy of the dispersion forces are considerably smaller in this region; in addition the contribution to the internal energy has the opposite sign. The values of the quadrupole moments determined from experimental sublimation energies, extrapolated to 0°K, are in good agreement with those calculated by other authors from microwave spectra. The difference in values of the quadrupole moments between nitrogen and carbon monoxide accounts for the anomaly in the densities of the α-modifi-cations: the density of α-CO is higher than that of α-N_2, although carbon monoxide has a larger "molecular volume" than nitrogen.

d) The effect of the permanent dipoles of carbon monoxide on the internal energy, specific heat, etc. is negligible at all densities up to the density of the crystal and at all temperatures.

APPENDIX I

It should be noted that the contribution to M_2 resulting from triplets of molecules is zero for orientational forces between permanent quadrupoles (the same is true for permanent dipoles), but that this term does not vanish for anisotropic London froces. It may be expected that this effect is charac-teristic for *induced* multipole interactions between nonspherical molecules. As a result, the quadrupole orientation energy decreases much more rapidly with increasing temperature than the interaction energy due to anisotropy in the dispersion forces. For a face centered cubic array the contribution of the triplet term in M_2 to the dispersion energy is about eight times larger than the contribution from pairs of molecules, at high temperatures.

It may also be expected that this effect is reflected in the results for the *third* virial coefficients of nitrogen and carbon monoxide gases. G u g g e n- h e i m (Disc. Far. Soc. **15**, (1953) 108, 109; Roy. Austral. Chem. inst. Rev. **3**, (1953) 1) remarks that the third virial coefficients of compressed nitrogen *cannot* be fitted by the same intermolecular parameters of a Lennard Jones potential as the second virial coefficients. Strictly speaking this criticism is only valid with respect to the use of a Lennard Jones potential (or any other spherically symmetrical two-body interaction) for the interpretation of physical properties of compressed gases consisting of *non*spherical molecules. It may well be that in this case the observed deviations for the third virial coefficients are due largely to the triplet terms in the anisotropic dispersion forces, discussed above.

Appendix II

The internal specific heats at constant volume of compressed nitrogen and carbon monoxide show a very peculiar behaviour. If the data on nitrogen are plotted versus temperature, a rather flat *maximum* occurs at a temperature of about $-35°C$ for densities above roughly 200 Amagat. The internal specific heat of carbon monoxide, on the other hand, varies only very little in the temperature region between 0° and 150°C. Note that the critical temperatures for these gases are -147 and $-140°C$, respectively. A similar maximum above the critical temperatures, again starting at densities of approximately 200 Amagat, has been found at the van der Waals laboratory for ethylene and air (it does *not* occur with carbon dioxide, however). It might be expected that orientational forces have something to do with these maxima. However, the terms in the internal energy which result from M_2 and M_3 and which are due to anisotropy in the London forces, *have the same sign*. Higher terms are too small to be of importance. This excludes the possibility for a maximum in the internal specific heats at constant volume due to orientational forces.

Unpublished results from the van der Waals Laboratory on the internal specific heat of compressed *xenon also show a maximum*; this occurs at roughly 65°C for all densities above approximately 200 Amagat. This also rules out a connection with orientation interactions.

REFERENCES

1) Part I, Chapter V of this thesis.
2) Part II, Chapter V of this thesis.
3) W e n t o r f, R. H., B u e h l e r, R. J., H i r s c h f e l d e r, J. O. and C u r t i s s, C. F., J. chem. Phys. **18** (1950) 1484.
4) S m i t h, W. V. and H o w a r d, R., Phys. Rev. **79** (1950) 132; H i l l, R. M. and S m i t h, W. V., Phys. Rev. **82** (1951) 451.
5) L e n n a r d-J o n e s, J. E. and I n g h a m, A. E., Proc. roy. Soc. A **107** (1925) 636.
6) G o e p p e r t-M a y e r, M. and K a n e, G., J. chem. Phys. **8** (1940) 642.
7) v a n I t t e r b e e k, and d e C l i p p e l i e r, Physica **14** (1948) 349.
8) M a r g e n a u, H., Rev. mod. Phys. **11** (1939) 1.

CHAPTER VI

SOME CONSIDERATIONS ON THE THEORY
OF TRANSITIONS IN MOLECULAR CRYSTALS

Many molecular crystals exhibit transitions of first or second order at low temperatures. The best known examples of such transitions are found with the crystals of the hydrogen and ammonium halides. In general the phenomena are extremely complex [1].

Most of the theoretical attempts to explain these transitions have been applied to this kind of crystals. Whereas P a u l i n g [2] and F o w l e r [3] explained these transitions in terms of the onset of free rotation of the molecules as temperature rises, it has later been found from observations on nuclear magnetic resonance [4] that in most cases the molecules do not rotate freely above the transition temperature. This corroborates F r e n k e l's [5] theory in which transitions were explained as changes in the orientational order of the molecules.

A second group of molecular crystals exhibiting transitions is formed by some diatomic molecules such as N_2, CO and O_2. It is the purpose of this chapter to outline a theoretical analysis of transition phenomena in crystals of this second group.

Nitrogen and carbon monoxide crystallize in two allotropic modifications, called α and β. The crystal structure of the α-form is face centered cubic (f.c.c.); the β-form is a hexagonal close packed lattice (h.c.p.) [6] [7]. The f.c.c. lattice is the stable structure at the lowest temperatures; at higher temperatures a transition occurs to the h.c.p. lattice.

As in the case of the hydrogen and ammonium halides, the transitions in the crystals of nitrogen and carbon monoxide are due to orientational effects. This follows from the fact that the heavy rare gases neon, argon, krypton and xenon, which also crystallize at the lowest temperature in the f.c.c. structure, do not exhibit a transition to the h.c.p. lattice. It has been shown that the orientational forces in the crystals of nitrogen and carbon monoxide at absolute zero constitute about 15% of the total crystal energy [8]. It can then be assumed that the spherically symmetric part of the interaction restricts the possible crystal structures of these diatomic molecules to either the cubic or hexagonal close packed lattice, but that the preference for one of these two structures is determined primarily by orientational forces. The most important deviations from spherical symmetry in the interaction field of crystalline nitrogen and carbon monoxide are caused by quadrupole orientation effects and by the anisotropy in the London forces [8]. The preferred orientations of the molecular axes are different for the two types of forces. As a result the orientational energy due to anisotropy in the London forces is even positive in the f.c.c. lattice at absolute zero, whereas the quadrupole interaction energy is strongly negative. On this basis it can be expected that the transition temperature increases with increasing values of the quadrupole moments. On the other hand, the correlation between transition temperature and anisotropy in the London forces should be the inverse.

In the following table the quadrupole moment Q, anisotropy in the polarizability γ, transition temperature $T_{tr(\alpha-\beta)}$ and transition heat ΔH are listed for crystals of nitrogen, carbon monoxide and oxygen.

TABLE I

Quantities related to transition phenomena in crystals of CO, N_2 and O_2				
	$\|Q \times 10^{+26} \text{e.s.u.}\|$	γ	$T_{tr(\alpha-\beta)}$ (°K)	ΔH cal/mol
CO	1.71	0.167	61.57	151.3
N_2	1.27	0.189	35.61	54.71
O_2	< 0.52	0.24	23.8	21

Oxygen has been included in the table because it shows the same trend as CO and N_2 with respect to the dependence of T_{tr} on Q and γ. However, it apparently does not crystallize in a f.c.c. lattice at the lowest temperatures (it may deviate too much from the ideal concept of a molecular crystal).

It seems plausible to assume that the analysis of these transitions may be based on an interaction potential between the molecules which consists of a spherically symmetric part, a term referring to quadrupole orientations and a contribution from anisotropy in the London forces. A "mixed" interaction has also been assumed e.g. by K r i e g e r and J a m e s [9] for molecules having a permanent dipole moment and anisotropic London forces. The effect of the interaction of permanent quadrupole moments has so far not been taken into account in the various theoretical treatments. With the above form of the molecular interaction the partition function of the crystal must be evaluated as a function of temperature. In general this problem is too complicated to be solved exactly; therefore an approximation method must be used. It is usually assumed that the analysis may be based on an orientation interaction depending only on the relative orientation of the molecular axes (and on the distance between the centers of the molecules). Further, to simplify the statistical problem, use is made of the following approximations (see e.g. ref. 9):

a. each molecule takes on its orientations in an average field of it neighbors;

b. the probability that a molecule i has its length axis oriented within the solid angle $d\omega_i$ about the direction (Θ_i, Φ_i) is $f(\cos \Theta_i)\, d\omega_i$; Θ_i is measured with respect to a fixed direction in space (determined by the crystal symmetry);

c. $f(\cos \Theta_i)$ is the same for all the molecules.

These three conditions are necessary and sufficient for the internal consistency of the analysis.

For the application of this method to the transitions in nitrogen and carbon monoxide the free energy of the f.c.c. and h.c.p. lattices must be evaluated separately as a function of temperature, because of the change in crystal structure accompanying the transitions.

An important feature in the present case is the fact that the orientational symmetry axes have different directions for different lattice sites in the f.c.c. structure. This implies that, even if the orientation interaction is the same for a specific microconfiguration of the ensemble in the two lattices, yet the probabilities of such a configuration may be different for the two structures, so that a transition from one structure to the other may in principle occur.

A detailed analysis of this problem is in progress.

REFERENCES

1) E u c k e n, A., Z. Elektrochem. **45** (1939) 126.
2) P a u l i n g, L., Phys. Rev. **36** (1940) 430.
3) F o w l e r, R. H., Proc. Roy. Soc. (London) A **149** (1935) 1.
4) A l p e r t, N. L., Phys. Rev. **75** (1949) 398; G u t o w s k y, H. S. and P a k e, G. E., J. Chem. Phys. **18** (1950) 162.
5) F r e n k e l, J., Acta Physiochem. U.R.S.S. **3** (1935) 23.
6) V e g a r d, L., Z. Physik **58** (1929) 496; **61** (1930) 185; **79** (1932) 471; **88** (1934) 235.
7) R u h e m a n n, M., Z. Physik **76** (1932) 368.
8) Chapter V of this thesis.
9) K r i e g e r T. J., a n d J a m e s, H. M., J. Chem. Phys. **22** (1954) 796.